Advanced Practical Chemistry

RESOURCE PACK

Edited by

Alec Thompson MA
Advisory Teacher, ILEA

Lambros Atteshlis BSc
Formerly Advisory Teacher
and member of the Science
Support Team, ILEA

Contents

Teachers' Notes
Technicians' Sheets
Master Sheets:
    Specimen Results and Answers to Questions
    Enlarged Results Tables

John Murray

© Inner London Education Authority, 1985

First published 1985
by John Murray (Publishers) Ltd
50 Albermarle Street, London W1X 4BD
Reprinted 1987, 1992, 1994

Printed in Great Britain by Athenaeum Press Ltd, Newcastle upon Tyne

British Library Cataloguing in Publication Data

Advanced practical chemistry resource pack.
  1. Chemistry
  I. Thompson, A.   II. Atteshlis, L.
  540 QD33

ISBN 0 7195 4255 3

TEACHERS' NOTES

for

Advanced Practical Chemistry

TEACHERS' NOTES FOR ADVANCED PRACTICAL CHEMISTRY

Experiment 1. Determining the Avogadro constant

Hair seems to give better results than cotton and does not need initial greasing.
It is important to make sure that the loop is in contact with the surface
throughout its circumference - otherwise the oleic acid will leak out.

The pupils should obtain a value for L to within one power of 10 of the accepted
value. You might ask more able students to investigate quantitatively the
effect on the calculation of making various assumptions about the shape of the
molecule.

Experiment 2. Preparing a standard solution

The students will need somewhere to keep their standard solutions until they do
Experiment 3 - perhaps a safe corner in the preparation room. They must check
that their solutions are clearly labelled and the labels firmly attached to
the flasks.

Experiment 3. An acid-base titration

In order to check the students' accuracy, either you or your technician should
standardize the sodium hydroxide solution used in the experiment. Since this
may be the students' first titration, you should watch them fairly closely and
discuss any errors in their technique.

Strictly speaking, we are not justified in quoting the final concentration in
our specimen calculation as 0.0994 mol dm^{-3} since this implies a precision of
1 part in 1000 whereas the concentration of the hydrogenphthalate solution
was quoted as 0.103 mol dm^{-3} (1 part in 100). However, the 'three significant
figures' approach has the merit of simplicity and is the approach we use in
most of our calculations.

Experiment 4. A redox titration

This is a straightforward experiment which should give good results. Freshly
prepared starch solution is required for a sharp end-point.

Experiment 5. A precipitation titration

The pupils are given a standard solution of silver nitrate. Since A.R. silver
nitrate has a purity of at least 99.9%, the technician can prepare a standard
solution by direct weighing. If, however, commercial recrystallized silver
nitrate is used, the prepared solution should be standardized either by you
or your technician with pure sodium chloride using potassium chromate
indicator.

Some students may experience difficulty in detecting the end-point of this
titration, i.e. the point at which the pale yellow colour just changes to a
faint reddish-brown colour. These students should be advised to prepare a
blank at the start of the experiment by mixing together about 10.0 cm^3 of the
barium chloride solution, 1 g of sodium sulphate, 5.0 cm^3 of the silver
nitrate solution and a few drops of indicator. The end-point of their titrations can then be taken to be the point at which the colour of their titration
mixture departs from the pale colour of the blank.

Warn your pupils not to waste or discard silver nitrate solution in view of its
prohibitive cost. See that they return the contents of their titration-
flasks to the silver residues bottle, so that the silver may be recovered and
silver nitrate regenerated.

Experiment 6. A titration exercise

Here is an opportunity to test your students' techniques and accuracy in
titration. If your particular syllabus requires practical work to be
internally assessed, then you could use this as one of your assessed
practicals.

In this practical, the students titrate ammonium iron(II) sulphate solution
against standard potassium manganate(VII) in order to determine the number of
molecules of water of hydration in ammonium iron(II) sulphate crystals, i.e.
determine x in the formula $FeSO_4(NH_4)_2SO_4 \cdot xH_2O$.

Neither potassium manganate(VII) nor ammonium iron(II) sulphate are primary
standards. However, it is sufficient for this experiment to prepare the
solutions by direct weighing provided that A.R. grade chemicals are used and
care is taken to avoid oxidation of the iron(II) by air.

Suggested mark-scheme

You should perform this titration yourself in order to set a standard against
which the students' titration values may be compared. We suggest you allot
two-thirds of your marks for accuracy and one-third for showing the correct
method in the calculation. A detailed discussion of mark-schemes appears in:

 Wellington & Doherty, Education in Chemistry, May 1981, 18, p82

Accuracy

In assessing accuracy, we suggest you give full marks for a titration value
which agrees with the standard (teacher's mean titration value) to ± 0.10 cm³,
with decreasing marks down to ± 0.30 cm³. No marks for accuracy should be
given for errors greater than ± 0.40 cm³. Some errors, of course, cannot be
blamed on the student.

Experiment 7. Estimating the ionization energy of a noble gas

This experiment is not of any great significance since ionization energies are
more properly discussed and measured by means of emission spectroscopy.
However, it is included to give students some practical activity in a theoretical topic and also to provide a direct introduction to the concept of
ionization energy.

A clearly marked circuit board, with valve already mounted, should be provided
so that students can connect the voltage supplies and instruments quickly and
correctly. This experiment should not become an exercise in practical
electronics! Also, you should try out the procedure yourself to make sure that
the particular equipment available to you does give a reasonable result - this
is even more important if you choose to adopt a different procedure (see below).

The simple method we describe can only give an approximate indication of
ionization energy, but we recommend that more sophisticated methods, which
involve plotting graphs of sets of readings, are not pursued by students except
perhaps as a special project. A survey of text-books, some of which mishandle
the topic, reveals a variety of different circuits that can be used and a
variety of possible curves relating, for instance, grid potential to anode
current. The exact nature of these curves seems to be dependent not only on
the circuit and method but also on the geometry of the particular valve used.
If you wish to pursue the matter further, we suggest the following references:

 Nyholm and Dureen, J.R.I.C., 1963, 87, p110
 Morton, School Science Review, 62, No. 218, Sept. 1980, p86

Experiment 8.   Using a hand spectroscope to observe the emission spectra of some *S*-block elements

Students should work in pairs because it is not easy to see many of the spectral lines in flame colours due to their low intensity and short duration. You may prefer, if it is possible, to borrow a set of discharge tubes and relevant apparatus from a physics department - the lines are much easier to see. In any case, you should make sure the students have access to some coloured plates, or transparencies, of emission spectra to help them to recognise the lines they see. A good plate is available from the Open University.

An alternative method of producing emission spectra, which can be made to work quite well, is to hold filter papers, soaked in the solutions, with tongs in a Bunsen flame.

Experiment 9.   Determining an enthalpy change of reaction

It is worth trying out this experiment to see that your sample of zinc reacts at a suitable rate. If the zinc is very finely powdered, or if it is in too great excess, the reaction occurs so rapidly that the point of plotting the temperature/time curve is lost. Conversely, coarse zinc in the stoichiometric amount may not react completely in a reasonable time so that the maximum temperature cannot be determined accurately.

Students should get excellent results - low results may be due to impure zinc (partially oxidized). Students may notice a side reaction, hydrogen being evolved from the acidic $CuSO_4$(aq), but this is small in extent and has a similar value for $\Delta H^{\ominus}$, so that little error is introduced.

Experiment 10.   Determining an enthalpy change of solution

This provides a good opportunity for a practical assessment on 'planning skills', as required by some examining boards. The experiment is very simple; requirements and procedure are similar to those in Experiment 9. However, a temperature/time curve is not necessary because the temperature change is much smaller and occurs very rapidly.

Since the temperature change is small, students should use a thermometer graduated in 0.1 °C. Suitable masses are 2.67 g of NH_4Cl (0.05 mol) and 90.0 g of H_2O - avoid too small a scale of operation. Students should observe a temperature drop of nearly 2 K, giving a value of $\Delta H^{\ominus}_{soln}$ within 2 or 3 kJ mol^{-1} of the accepted value of 16.4 kJ mol^{-1}.

Experiment 11.   Using Hess's law

The choice of magnesium sulphate over the more commonly used copper sulphate enables good results to be obtained in simple apparatus because the temperature changes are greater and the salt more readily dehydrated. However, if you wish your students to use a more sophisticated apparatus and method, you can give them Experiment 12 as an alternative. In both experiments, some care should be taken to obtain, and keep dry, samples of the anhydrous salts. A sensitive thermometer is essential.

Experiment 12. Another application of Hess's law

The experimental procedure is straightforward but it takes longer than Experiment 11 and the calculation is more difficult - we suggest that weaker students do not attempt it.

It may be worth purchasing a small amount of anhydrous copper sulphate each year since it is not too easy to dehydrate the crystals properly (see Technician's sheet). Poor purity of the anhydrous salt, and its very slow rate of dissolving, could account for a smaller temperature rise than the 6.5 K needed to obtain a good result. Although the temperature change on dissolving the hydrate is much smaller (-1.2 K), students usually obtain a better result here.

Experiment 13. Determining heats of combustion

This experiment is lengthy and requires good experimental technique, but is well worth doing if time permits. Show one of the two ILPAC video programmes entitled 'Using a heat of combustion apparatus' beforehand - the first refers to the glass envelope/copper spiral apparatus illustrated in the Unit, while the second refers to an improved all-metal apparatus (supplied by Philip Harris), which is recommended although slightly more expensive. Alternatively, you may wish to use the home-made apparatus described in Appendix 2 of the Nuffield Advanced Chemistry Teachers' Guide II - used with care, this can give quite good results. Whichever apparatus is used, it is well worthwhile obtaining 6 or 7 small spirit lamps - they can be purchased separately from Griffin. This will eliminate errors arising from spillage and cleaning, and will save a good deal of time.

Some difficulty may be experienced in keeping the lamps burning satisfactorily during the experiment, particularly if the water pressure is such that filter pumps do not work very well. It is worth cleaning scaled-up tap apertures which can cut down water flow considerably. Standing the lamps on small blocks improves air flow but it is advisable to check quickly that the lamp will burn properly in position under the calorimeter before the temperature and weight are recorded.

It is important that for each determination the calorimeter should be refilled (to the mark) with fresh water from a supply which has stood in the laboratory overnight. Successive samples of tap water can vary significantly in temperature and this can invalidate the calibration of the apparatus. It is also important for the calorimeter to be used at exactly the same height for each determination.

Results tend to be low largely due to incomplete combustion but careful work should give a fairly constant difference between values for successive alcohols.

Experiment 14. A thermometric titration

This experiment is straightforward and should give results within 2 or 3 kJ mol^{-1} of the data book values. Unfortunately, experimental error is sometimes more significant than the difference between the two values, but usually both results are low by about the same amount for the usual reasons - heat loss and assumptions about the heat capacities of the solutions.

The concentration of the sodium hydroxide solution must be precisely known - it is worth purchasing some ampoules of concentrated solution for making it up. It may be worth checking the concentrations of the acids to make sure that the equivalence point is not precisely at a multiple of 5 cm^3 of acid. This will ensure that students extrapolate their graphs properly.

Experiment 15. Making models of two metallic structures

Any size of expanded polystyrene sphere can be used, but the larger the better for ease of handling.

Blu-tak seems to work rather better than Buddies.

Students should be encouraged to use adhesive on _all_ the contacts mentioned in the instructions and to press spheres as close together as possible. In this way remarkably strong models can be assembled.

Experiment 16. Recognising ionic, covalent and metallic structures

Students should be discouraged from guessing the _identity_ of the substances - they should be thinking about the relationships between _properties_ and _structure_.

It might be worth disguising the white solids by grinding them to fine powders and mixing with a _little_ carbon so that they all look grey. Students should also be encouraged to _select_ tests rather than carry out five tests on each of the seven substances, and to be critical about the interpretations of their results.

If your students have not come across naphthalene in their pre-A-level course, you may wish to substitute it for glucose, D.

Experiment 17. Determining the molar mass of a gas

You may prefer to use a syringe method (see Nuffield, Topic 3), but we have found that students do not readily understand the buoyancy correction and, in any case, get poorer results using a syringe.

If you allow students to operate a gas cylinder, be sure to instruct them carefully - we suggest that they use a needle valve _only_, after you have opened the main valve and set the regulator.

If you use a gas generator, e.g. Kipp's, you should instruct the students (and/or technician) how to purify the gas before use.

Some students may realise that the densities given refer to _dry_ air. However, this makes little difference to the results, which can be very good with this simple method.

(The treatment of results does not explore the effect of Archimedean upthrust on the empty flask, as pupils appear to find this distracting. In fact, the calculation implies that the volume of air inside the flask is equal to that of the air displaced by the flask, a not unreasonable assumption.)

Experiment 18. Determining the molar mass of a volatile liquid

This experiment also gives good results but requires more manipulative skill, especially in transferring liquid to the gas syringe without loss.

Make sure that steam generators are fitted with a safety tube and that students connect them properly to avoid scalding.

Experiment 19. Determining the molar mass of a gas by effusion

This experiment gives good reproducible results if the apparatus is set up carefully. You may find that students have wrist-watches with stopwatch facilities more accurate than laboratory stopclocks - reading to 0.1 sec appears to be justified.

If you allow students to operate a gas cylinder, be sure to instruct them carefully and supervise closely - we suggest that they use a needle valve only, after you have opened the main valve and set the regulator. Also ensure that no flames are in the laboratory.

During the course of the experiment, you might check that students understand the difference between diffusion and effusion, and that Graham's law applies to both. We have assumed some knowledge of simple diffusion experiments.

We do not think it necessary to explain why the pin-hole should be small, other than to obtain measurable rates of effusion, but you may wish to discuss this with more able students.

Experiment 20. The effect of concentration changes on equilibria

This is a very simple experiment and should take only a few minutes. We suggest that students should not get involved at this stage in discussing the various possible formulae of the complex ion(s) formed by iron(III) in the presence of excess chloride ions. For this reason, we have written an equation in the simplest possible form.

Experiment 21. Determining an equilibrium constant

This experiment can give excellent results if the procedure is followed carefully. However, students may need help in understanding the procedure if it is not to become a mechanical exercise. Weaker students especially will need guidance through the rather lengthy calculations. If you have a suitable water bath, you may wish to introduce temperature control, but we do not specify this because the equilibrium constant varies very little over the normal range of room temperatures. Carefully standardized sodium hydroxide solution is needed for this experiment.

Experiment 22. Determining a solubility product

Calcium hydroxide is too soluble for the solubility product, in terms of concentrations rather than activities, to be truly constant, and this is why some data books do not list a value. However, a simple titration method cannot be readily applied to less soluble salts, and this experiment does give quite good results.

Experiment 23. Illustrating the common ion effect

This is a simple experiment and should take only a few minutes. Note that a soluble salt such as sodium chloride can be used to illustrate the common ion effect even though the solubility product principle cannot be applied (in terms of concentrations) because of the interaction between ions at high concentrations.

Experiment 24. Distribution equilibrium

The use of trichloromethane (chloroform) or tetrachloromethane (carbon
tetrachloride) is now discouraged; this is why we have specified
1,1,1-trichloroethane as the second solvent even though ammonia is much less
soluble in this than in the other two solvents mentioned. Although we have
recommended that both hydrochloric acid solutions should be standardized,
you may feel that it is sufficient to make up approximately 0.5 M HCl and
then dilute this carefully so that the _ratio_ of concentrations is precisely
50:1. The end-point may not be very sharp using 0.010 M HCl; you may wish
to warn your students of this.

Note that the use of another indicator (e.g. methyl red) may cause problems if
either the acid form or the base form is molecular and, therefore, so soluble
in the organic solvent as to leave a concentration in the aqueous layer which
is too small to give a distinct colour. Methyl orange is particularly suitable
because both forms are ionic.

Experiment 25. The pH of a weak acid at various concentrations
Experiment 26. The pH of different acids at the same concentration

These experiments take only a few minutes, and one pH meter can easily serve
a class. We suggest that, at this stage, you teach your students to use a
pH meter as a tool without considering how it works; a brief account can
be given, if required, during the study of redox reactions.

We think it best not to measure the pH of hydrochloric acid for comparison
purposes; meters actually measure -log (activity) and, consequently,
0.1 M HCl appears to have a pH greater than 1.

If a pH meter is not available, you may wish to try using narrow-range pH
papers, but they are not very satisfactory over this range.

Experiment 27. The action of a buffer solution

Students may not observe pH changes equal to those we have quoted in our
specimen results, but the contrast between the buffer and water should be
quite clear. If you use commercial buffer tablets, we suggest you check that
the solution is concentrated enough to absorb the amounts of acid and alkali
we have specified - it may be necessary to dilute the NaOH and HCl solutions.
The buffer recipe we have quoted contains more HPO_4^{2-} than $H_2PO_4^-$;
consequently, it absorbs acid more effectively than alkali.

Students may have some difficulty in measuring the pH of pure water quickly
enough before absorption of carbon dioxide decreases the value.

Experiment 28. Preparation of buffers: testing their buffering
capacity and the effect of dilution

You may choose to use this experiment to test planning and recording skills
as part of a practical assessment. Students may need help in calculating
the volumes of solutions to use and in understanding why the pH of the
buffer does change a little on dilution when the equation they have used
suggests that the pH should remain constant. The solutions we have specified
give concentrated buffers with a very high buffering capacity; you may
choose to use 0.1 M solutions for reasons of economy.

Experiment 29. Determining the pH range of some acid-base indicators

We suggest that results are shared amongst class members so that individual students do not spend too much time on this experiment. Make sure that students understand the procedure, particularly the need to have the same concentration of indicator in the solutions being compared.

The Universal Buffer solution supplied by BDH contains phenylethanoic acid, which has a most unpleasant smell! Students should be warned to keep solutions stoppered or covered and to neutralize waste with an alkali. The alternative contains ethanoic acid instead and, although its initial pH is slightly higher (3.25), the approximate relationship between pH and volume of added alkali still holds.

Experiment 30. Determining the ionization constant for an indicator

Bjerrum wedges are expensive to buy, but satisfactory ones can be made from scraps of perspex. The wedge method should give somewhat better results than the tube method because it gives a continuous range of colours. It also avoids errors due to the lens effect of curved tubes and variable air spaces. It is more attractive to students, but more expensive in materials consumed. Difficulty in colour matching is often due to varying concentrations of indicator - the volumes used should be carefully matched.

A ratio of 2:1 for $[\text{HIn(aq)}]:[\text{In}^-\text{(aq)}]$ gives pK_{In} = 4.0. Students should obtain a good result using either method because ratios as different as 3:1 and 1:1 give pK_{In} values of 4.2 and 3.7 respectively.

Experiment 31. Determining the dissociation constant of a weak acid using an indicator

The procedure is very similar to that for Experiment 30, and need not be attempted if time is short. Again, students should obtain good results using either method.

Experiment 32. Obtaining pH curves for acid-alkali titrations

The experiment is much simpler to do, and gives better results, if precisely 0.100 M solutions of acids and bases are provided (see Technicians' sheet). If this is not possible, the concentrations should be either precisely equal (but not necessarily precisely known, as long as they are near 0.1 M) or precisely known (so that adjustments to the results tables can be made).

If the concentrations are not equal, the volume, V cm^3, of alkali required to reach the equivalence point should first be determined and then the pH measured at V ± 0.05, 0.1, 0.2, 0.4, 0.6 cm^3 instead of the central values given in the Results Tables.

Experiment 33. The variation of boiling-point with composition for a binary liquid mixture

This experiment should, ideally, be done in a fume cupboard. Students should be closely supervised to see that they handle flammable liquids with care and have minimum exposure to trichloromethane vapour. We suggest that each student, or pair, works with only one mixture in order to save time and materials. Make sure that the technician disposes of residues carefully see Technicians' Sheet for Experiment 34.

Experiment 34. Measuring temperature changes on forming solutions

This is a very simple experiment and takes only a few minutes. We have omitted the calculation of values of ΔH_{mix} so that students focus on qualitative aspects of the experiment. However, we suggest that you remind students of the calculation - they will need to do a similar calculation in the next experiment.

Experiment 35. Determining the approximate strength of a hydrogen bond

We have left students to plan their own procedure for this experiment, based on their experience of Experiment 34, so that you can assess them on this skill (and others) if you wish. Some students will need help in understanding that an excess of one liquid should be used so that the other liquid is as completely hydrogen-bonded as possible. 5:1 or 6:1 is probably sufficient, but our specimen results (below) are based on a 10:1 mixture. A ratio greater than this gives too small a temperature rise. Students may also need help with the calculations.

Students will need the specific heat capacities of glass and of the two liquids.

Glass:	0.67 J g^{-1} K^{-1}
Trichloromethane:	0.96 J g^{-1} K^{-1}
Methyl ethanoate:	1.97 J g^{-1} K^{-1}

Experiment 36. The effect of hydrogen bonding on liquid flow

It is worth keeping sealed tubes of these four liquids permanently in store; no further preparation is needed and the experiment takes only a few minutes.

Experiment 37. Testing liquids for polarity

To minimize inhalation of vapours, corked burettes containing the liquids should be supplied for the students and the experiment should be done in a fume cupboard.

It is difficult to standardize the procedure but the range of deflections should be broadly in line with the dipole moments of the liquids.

Experiment 38. Determining relative molecular mass by a freezing-point method

This experiment is straightforward and should give good results if the instructions are followed carefully. You may think it worthwhile to ask students to calibrate the thermometer by measuring the melting-point of pure A, but this procedure was not specified by the examiners.

Experiment 39. Determining enthalpy changes and volume changes of solution

The first part of this experiment, determining enthalpies of solution, is straightforward provided that samples of the salts are dry. Anhydrous iron(III) chloride is difficult to obtain and store - it may be worth buying sufficient for a demonstration only. Expanded polystyrene cups have negligible heat capacity and their use makes the calculations much simpler.

The second part requires more care. We suggest that you supervise the experiment, particularly if anhydrous iron(III) chloride is used, and ensure that students understand the procedure for measuring the volume change. If gas burettes are available, the procedure can be simplified since there is no uncalibrated 'dead space'.

Note that, although exothermic solution is usually accompanied by a volume reduction, the correlation between the two effects is not entirely clear and students would not be expected to explain the anomalies.

Experiment 40. Investigating the hydrolysis of benzenediazonium chloride

This experiment is more straightforward than it appears at first sight. Make sure that your students do not start to collect the gas until the reaction mixture has reached a steady temperature, and that they understand why it does not affect the results to allow gas to escape before this point is reached. Students may also need help with the use of $V_\infty - V_t$ as a measure of the concentration of benzenediazonium chloride.

Better results are obtained if the reaction mixture is <u>continuously</u> stirred, but this is not practicable in most school laboratories. <u>Occasional</u> stirring should be avoided, since this results in uneven evolution of gas.

Check that the syringe piston moves freely and encourage your students to rotate it in order to equalise pressures before taking a volume reading. Also, make sure that they know in advance what to do when the gas volume exceeds the volume of the syringe.

Experiment 41. The kinetics of the reaction between iodine and propanone in aqueous solution

This experiment often causes problems because of the unreliability of inexpensive colorimeters. We suggest that you try out the procedure in advance in order to decide whether the variable light output of the lamp is more or less important than the precise positioning of the tube (see also our answer to Question 5). Optically matched tubes are preferable, but adequate results may be obtained with a carefully selected set of ordinary tubes, or even by inserting the same tube into the colorimeter for all readings. All tubes must be wiped clean at the lower end and handled only at the top.

Another problem is that heat from the bulb may increase the temperature of the mixture by several degrees over six minutes. However, this is not as serious as it seems, because it is the <u>initial</u> rate that is required. Calculations show that, for the initial concentrations we specify, each graph should be a virtually straight line; a temperature rise gives a curve showing the rate <u>increasing</u> slightly (as in our specimen results), but this has the same slope at $t = 0$. Note that the autocatalytic effect of increasing $[H^+(aq)]$ is not evident under these conditions because $[CH_3COCH_3(aq)]$ decreases as $[H^+(aq)]$ increases, and neither of these concentrations changes very much during the experiment. (Continued on next page.)

(Continued from last page.)

Our results were obtained using a WPA CO65 colorimeter. Other colorimeters
may give better results over a different range of iodine concentrations. If,
for instance, a calibration curve suggests that the best range is from 0 to
0.005 mol dm^{-3}, you could use 0.010 M I_2 instead of 0.020 M I_2. Nothing else
need be changed, because the rate of reaction is independent of the concentration of iodine.

Students sometimes obtain a series of values of $[I_2(aq)]$ which give a good
line only if the calculated value at t = 0 is ignored. This is probably due
to a 'zero drift' on the instrument. In this event, the slope of the line at
t = 30 s should be taken.

If you are not satisfied that your colorimeter(s) can give satisfactory results,
your students could study the same reaction by a titration method (see, for
instance, Nuffield Advanced Chemistry, Topic 14) or they could study a
different reaction by a 'clock' method (see Experiment 44).

Experiment 42. Determining the activation energy of a reaction

This experiment is straightforward and should give good results if the
procedure is followed carefully. Make sure your students understand that
they need not waste time adjusting the temperature precisely to the values we
have stated, but that they must record the actual temperatures carefully.
Note, however, that at temperatures higher than 60 °C, the starch-iodine
complex is broken down (reversibly) and the blue colour does not appear!

You may wish to point out to your students that plotting log t instead of
log($1/t$) is simpler and only changes the sign of the slope.

Experiment 43. Determining the activation energy of a
catalysed reaction

You can use this experiment for the assessment of planning abilities, if you
wish. Students would be expected to base their plan on the procedure for
Experiment 42 but work at lower temperatures. You may wish to suggest using
lower concentrations, however, because those specified for Experiment 42 give
rather short reaction times - see our specimen results.

Experiment 44. A bromine 'clock' reaction

This experiment is more straightforward than it appears and should give good
results provided students prepare their reaction mixtures carefully and mix
them thoroughly. Some care is needed in the preparation of solutions by the
technician - in particular, the phenol solution is so dilute that its concentration cannot be relied upon unless it is made up just before the experiment
by dilution of a more concentrated stock solution. Students may pool their
results to save time - temperature variation between sets of results does
not matter as long as it is constant within each set.

Experiment 45. Some simple redox reactions

You may feel your students are already familiar with this simple experiment
and may omit it, but we suggest that they do answer the questions. Remind
your students not to wash away solutions containing silver salts - this
applies to Experiments 46, 48 and 49 as well.

Experiment 46. Measuring the potential difference generated by some simple electrochemical cells

This is a simple experiment but it is most important for what follows that your students understand how the polarity of the electrodes enables them to determine what reaction is occurring.

Experiment 47. Testing predictions about redox reactions

You may wish your students to make predictions _before_ they do the experiment rather than answer the question afterwards. Reaction F does not occur as . expected, presumably because of a kinetic effect (high activation energy). This limitation on predictions from $E^{\ominus}$ values will be mentioned again but the point cannot be made too often.

Experiment 48. Variation of cell potential with concentration

The silver nitrate solutions should be made up accurately and students encouraged to work carefully so that a good graph of E/V against $\log[Ag^+(aq)]$ can be drawn for use in following exercises. The straight-line plot has to be extrapolated to $\log[Ag^+(aq)] = -15$, which can introduce considerable error; your students may need help with this.

Students should work from the most dilute solution to the most concentrated, so that accidental transfer from one beaker to the next causes negligible change in concentration. Salt bridges should be drained before use and left in the solutions for the minimum time in order to reduce contamination. If this procedure is followed, the same solutions may be used by many students.

Experiment 49. Measuring the solubility products of some silver salts

Ensure that your students revise the concept of solubility product before they do this experiment, which provides an alternative to the titration method used for calcium hydroxide in Experiment 22. The calculation involves using the graph obtained from Experiment 48; to emphasise the limited accuracy of the results, you might suggest that students make two calculations for each silver salt, using two lines on the graph to represent the upper and lower limits of error in extrapolation.

Experiment 50. Reaction between sodium peroxide and water

Sodium peroxide must be reasonably fresh - it absorbs carbon dioxide from the air, turning from pale yellow to white in the process. You may be more familiar with a test for hydrogen peroxide using ethoxyethane (diethyl ether), but pentan-1-ol is safer and gives the same result provided the procedure is carefully followed. Ethyl ethanoate may be used instead of pentan-1-ol.

Experiment 51. Heating the nitrates and carbonates of the S-block elements

This is a very simple experiment and should not be allowed to take up too much laboratory time, particularly if students are familiar with these reactions from pre-A-level work. Encourage students to see the importance of following the same procedure as precisely as possible for successive carbonates or nitrates. Decomposition times may vary widely from our specimen results but should show the same trends. We have found that potassium and sodium carbonates seem to decompose more readily than text-books suggest.

Experiment 52.   The solubility of some salts of Group II elements

This experiment works well provided the solutions are of the right concentrations, and the trends in solubility should be readily apparent. The sodium hydroxide solution should be fresh since contamination with carbonate can cause unwanted precipitates.

Experiment 53.   The solubility of the halogens in organic solvents

Marked colour changes indicate that bromine and iodine are more soluble in these organic solvents than in water. Chlorine is also more soluble in organic solvents than in water, but this cannot be deduced directly from the experiments because colour changes are scarcely noticeable. Note that the colour of iodine in 1,1,1-trichloroethane varies from violet (dilute) to red (concentrated).

Cyclohexane may be used instead of hexane and gives the same results. This experiment also provides an opportunity to train your students in the proper disposal of organic solvents (see note before Experiment 81).

Experiment 54.   The action of dilute alkali on the halogens

This is a very straightforward experiment and should take only a few minutes to perform. However, make sure your students use their textbooks to answer the questions properly.

Experiment 55.   Halogen-halide reactions in aqueous solution

If your students have done this experiment in their pre-A-level course, you may feel they need not repeat it here. However, make sure they understand the use of an organic solvent for identification, especially to distinguish between bromine and iodine. Note that the use of trichloromethane (chloroform) and tetrachloromethane (carbon tetrachloride) is now strongly discouraged where suitable substitutes are readily available; 1,1,1-trichloroethane is not a good substitute in this experiment because the colour of solutions of iodine varies considerably with concentration. Cyclohexane, which is cheaper, may be used instead of hexane. You may also wish to introduce the use of starch solution as a test for iodine.

Experiment 56.   Reactions of solid halides

Make sure your students take care in using concentrated sulphuric acid and know how to deal with minor spillages. Orthophosphoric(V) acid should also be treated with care. Some success may be obtained using 'syrupy' phosphoric acid, containing about 10% of water, but 100% phosphoric acid, which is a solid, is better. Syrupy phosphoric(V) acid can be dehydrated by cautiously adding phosphorus(V) oxide, P_4O_{10}.

Experiment 57.   Reactions of halides in solution

Your students may have done some of these test-tube experiments before, but they are probably worth repeating quickly. You may choose to introduce the formulae of complex ions, such as $[Ag(NH_3)_2]^+$, to explain the solubility of silver halides in ammonia solution, but a detailed study is best left till later.

Experiment 58. Reactions of the halates

This experiment is fairly straightforward and should show clear differences
between the halates, e.g. in oxidizing power. Although iodate(V) and iodide
do not react in neutral solution, there may be sufficient hydrogen ions in
distilled water to give a trace of iodine. Note that the chlorine released
from a solution of chlorate(I) by the addition of acid comes partly from Cl^-
and partly from ClO^- - a sort of 'reverse disproportionation'.

Experiment 59. Balancing a redox reaction

Encourage your students to polish their titration technique so that they
obtain good agreement in consecutive titres, even though great accuracy is
not strictly necessary in this experiment (the answer has to be an integer,
and an error of as much as $\pm$ 0.5 cm^3 in the titre still leaves it fairly
clear that the integer is 5). Make sure the starch solution is freshly
prepared so that a clear end-point can be observed.

Experiments 60 & 61. Observation and deduction experiments

Most students need a lot of help when first performing experiments of this
sort, in making observations, in interpreting them and in reporting <u>fully</u>
what they see and deduce. We suggest that you spend as much time as possible
discussing the experiments with individual students while they are performing
them. Make sure your students realise that, in an examination, many marks
can be obtained for detailed and accurate reporting of observations, even
when no firm conclusions are drawn from them.

Experiment 62. Investigating the properties of Period 3 chlorides

In this experiment we have asked students to use anhydrous aluminium chloride
and hydrated magnesium chloride. You may feel that this is an unfair
comparison and decide to use anhydrous magnesium chloride which is available
cheaply as a technical grade reagent. This solid gives a temperature rise
of about 7 K when added to water but the resulting solution is strongly
alkaline. We have therefore decided that the technical grade is too impure
for the purposes of this experiment. Hydrated aluminium chloride gives a
very small temperature rise when added to water (about 1 K) and the
resulting solution is strongly acidic. This could be used in place of the
anhydrous aluminium chloride.

Silicon tetrachloride is readily hydrolysed and the product can firmly
seal a stopper in place and allow pressure to build up in the bottle.
Bottles containing silicon tetrachloride have been known to burst in store,
or on being opened. Bottles of $SiCl_4$ should be opened in the fume cupboard
with considerable caution after covering the bottle with a cloth. It is
particularly dangerous to return unused chemical to the bottle.

Cyclohexane may be used instead of hexane and gives similar results. When
disposing of organic residues ask your students to avoid pouring undissolved
solid into the residues bottle.

Experiment 63.   Preparing anhydrous aluminium chloride

This experiment requires a fair degree of manipulative skill. A helping
hand from you will be particularly useful to the student. Several accidents
have been reported when preparing chlorine from concentrated hydrochloric
acid and potassium manganate(VII) (permanganate), when concentrated sulphuric
acid has been used in place of hydrochloric, with a consequential explosion.
Concentrated sulphuric acid should not be in the laboratory when preparing
chlorine in this way, and must certainly not be used to dry the gas.

Experiment 64.   Investigating the properties of Period 3 oxides

This experiment is fairly straightforward and similar to Experiment 62.

When bubbling SO_2 through a measuring cylinder containing water, the
students will see a significant decrease in the size of the gas bubbles as
they rise to the surface of the water. The bubbles may dissolve completely
before they reach the top.

Experiment 65.   The reactions of tin and lead and their
aqueous ions

This is a fairly straightforward experiment and works well. Sodium sulphide
has been used in place of the more hazardous hydrogen sulphide and gives
similar results. Old samples of tin(II) chloride are very difficult to
dissolve, even in concentrated hydrochloric acid. If new stock is not
available, then an adequate solution can be made by reacting tin with concentrated hydrochloric acid, followed by dilution of the resulting solution.

Experiment 66.   The preparations and reactions of tin(IV) oxide
and lead(IV) oxide

In the preparation of lead(IV) oxide, students should ensure that the product
is completely dry, otherwise they will end up with apparent percentage yields
in excess of 100%. Since lead(IV) oxide is easily decomposed by heat, we
suggest you dry the product in an incubator (borrowed from the biology
department) set at about 35 °C, and left for one or two days. Conventional
ovens tend to reach much higher temperatures even on the lowest settings.
Tin(IV) oxide seems to be particularly 'passive' when prepared in this
manner. It is therefore advisable to use samples from the manufacturers for
part C of the experiment. It is very difficult to detect whether the oxides
dissolve in acids and alkalis by this method and you may need to tell your
students that text-book(s) claim that they do.

Experiment 67.   Observation and deduction exercise

If your students have attempted other observation and deduction experiments
in the students' book, they will find this one relatively easy. Remind them
that, in an examination, many marks can be obtained for detailed and accurate
reporting of observations, even when no firm conclusions are drawn from them.

Experiment 68. Illustrating the oxidation states of vanadium

Zinc dust reduces vanadium(V) more rapidly than does granulated zinc and also
seems to give less hydrogen (or perhaps the hydrogen is purer and therefore
smells less!).

Campden tablets (used for sterilizing wine-making equipment and consisting
largely of sodium disulphate(IV) - sodium metabisulphite - $Na_2S_2O_5$) may be
used to reduce V(V) to V(IV).

Experiment 69. Illustrating the oxidation states of manganese

If time is short, this experiment could be omitted.

The alternative method mentioned for making Mn(V) involves reduction of
Mn(VII) with 12 M KOH. In view of the hazards of handling very concentrated
alkali, this is suggested as a teacher demonstration only. Hydroxide ions
are oxidized to oxygen. Details can be found in Nuffield Advanced Science:
Chemistry, Teacher's Guide II, Section 16.1 (First Edition).

Experiment 70. Relative stabilities of some complex ions

Other ligands may be used if necessary - see, for instance, Nuffield Advanced
Science: Chemistry (First Edition), Section 16.2. Note that the addition of
sodium ethanedioate gives a precipitate initially, which dissolves in excess.

Experiment 71. Determining the formula of a complex ion

The experiment was tested using a WPA colorimeter (CO65). The concentrations
given, and some of our instructions, might not necessarily be suitable for
other colorimeters.

Adequate results can be obtained using no filters, but we think that the
principles of filter selection are important enough to be included.

Experiment 72. Some redox chemistry of copper

You may feel that this experiment is not essential, especially if copper(I)
compounds were studied at O-level.

Note that the use of boiling concentrated hydrochloric acid, often recommended
for the preparation of copper(I) chloride, is not necessary. If copper(II)
chloride is not available, a mixture of copper(II) sulphate and sodium chloride
may be used.

Experiment 73. Investigating the use of cobalt(II) ions as a catalyst

Experiment 74. Catalysing the reaction between iodide ions and peroxodisulphate

You may feel that students need not attempt both of these experiments.
Experiment 73 is much simpler and shorter, but the chemistry in Experiment 74
is probably more important.

Experiment 75. Observation and deduction exercise

Students may ask about the reaction in 4(c) for which no inference was
required. The transient deep-blue colour is due to the unstable peroxide
which is usually given the formula CrO_5. Students may recall seeing this
before in a test for hydrogen peroxide - see Experiment 50. The net result of
the reaction is the reduction of Cr(VI) to Cr(III) and the oxidation of
hydrogen peroxide to water and oxygen, as might be predicted from standard
redox potentials.

Experiment 76. Reactions of aluminium

Since mercury(II) chloride is toxic, we have restricted its use to one or two
drops. Copper(II) chloride removes the oxide layer just as effectively for
reactions in solution but does not promote a reaction with oxygen, presumably
because the oxide layer is re-formed. Note that chloride ions seem to be
necessary for the removal of the oxide layer - other mercury(II) or copper(II)
solutions are not effective.

Experiment 77. Anodizing aluminium

If you decide to omit this experiment to save time, we suggest that students
should read the instructions so that they know how it is done. A larger
cathode shaped as a cylinder may be used to surround the anode but the
process is well illustrated with the simple apparatus we have described.

Experiment 78. Reactions of the oxo-acids of nitrogen and
their salts

Ideally, the whole experiment should be performed at a fume cupboard. If
space is limited, some of the tests may be done in a well ventilated
laboratory but test 1 must be done at a fume cupboard and on a small scale
since NO_2 is toxic.

Experiment 79. Investigating some reactions of the oxo-salts
of sulphur

We have included many more reactions than students would be expected to know
and explain; we suggest that you emphasise that they are not expected to
learn all the equations! However, these reactions provide a very useful
exercise in the accurate recording of observations.

The silver nitrate tests may be omitted if you feel they are too extravagant
- they are unlikely to appear in practical examinations but students should
know about the dithiosulphatoargentate(I) complex in connection with its use
in photography.

Note that the conversion of sulphite to thiosulphate is far from complete,
so that X still gives the reactions of sulphite ions.

Experiment 80. Observation and deduction exercise

Remind your students to look at the wording of the question - in this
case comments on the types of reaction are required.

Notes on experiments in organic chemistry

Before the students start the experiments in this section, we recommend a group
discussion on the hazards of using organic compounds. Since the list of
dangerous or suspect chemicals is growing all the time, we think that
students should be advised to treat all organic compounds as potentially
dangerous, even though this may mean treating some chemicals with more care
than is necessary. We suggest the following guidelines.

Many organic compounds are:

(1) flammable, so ALWAYS STOPPER THE BOTTLES AFTER USE AND REMOVE THEM TO A
     SAFE PLACE AWAY FROM FLAMES;

(2) toxic if inhaled, so ALWAYS HANDLE ORGANIC CHEMICALS IN A FUME CUPBOARD
     (with the extractor switched on, of course);

(3) toxic if absorbed through the skin even though there may be no
     immediate effect, so ALWAYS WEAR GLOVES;

(4) violent in some of their reactions, so ALWAYS WEAR SAFETY SPECTACLES
     AND PULL DOWN THE FRONT OF THE FUME CUPBOARD (leaving a gap just large
     enough for the extractor to work efficiently).

The School Science Service (formerly the C.L.E.A.P.S.E. Development Group) at
Brunel University have printed a set of HAZCARDS which not only state the
potential dangers of many chemicals but also tell you what to do in the event
of an accident. It is a good idea for the technician to issue a HAZCARD with
each chemical. HAZCARDS are available from:

          School Science Service,
          Brunel University,
          Uxbridge.   UB8 3PH

          Telephone: Uxbridge 51496

A qualified science teacher should always be in a laboratory throughout
these experiments.

Disposal of organic residues

In certain circumstances, it may be acceptable to pour small amounts down a
sink in a fume cupboard with plenty of flowing water. However, it is better
to collect residues and dispose of them quickly by burning in a metal tray
in the open away from people and buildings. Alternatively, they could be
allowed to evaporate from a metal tray covered with a mesh on a roof with
controlled access. Normally, however, residues should not be stored nor
should residues from different experiments, containing different chemicals,
be mixed. For disposal of larger quantities, consult your local authority.

Further details can be found in laboratory manuals such as:

     Hazardous Chemicals - a Manual for Schools and Colleges - SSSERC
     published by Oliver and Boyd. ISBN 0 05 003204 6

Experiment 81. Chemical properties of alkanes

The only positive reaction, apart from combustion, in this experiment is the
reaction between cyclohexane and bromine (inert solvent) in the presence of
ultraviolet light. By blowing a little ammonia across the top of the tube,
students should see white fumes of ammonium bromide.

The action of bromine water may confuse students. The bromine tends to come
out of the water and dissolve in the cyclohexane - this is not an indication
that a reaction has taken place.

Experiment 82.   Chemical properties of alkenes

1,1,1-trichloroethane is chosen as the inert solvent in which to dissolve
bromine. This replaces tetrachloromethane and trichloromethane, which are
now considered to be more harmful.

The reaction of ethene with bromine in an inert solvent is now banned as a
preparation since the product, 1,2-dibromoethane, is carcinogenic; it is
wise, therefore, to avoid testing ethene in this way. In the absence of any
evidence about the possible dangers of 1,2,dibromocyclohexane, students should
err on the side of caution and dispose of the product quickly.

Experiment 83.   Hydrolysing organic halogen compounds

The order of ease of hydrolysis in this experiment is 1-iodobutane >
1-bromobutane > 1-chlorobutane > chlorobenzene.

Ethanol is added in this experiment to prevent the halogeno-compounds forming
an emulsion with the water present, since this can be confused with the
formation of a precipitate of silver halide.

Experiment 84.   Preparing a halogeno-alkane

We have chosen the preparation of 2-chloro-2-methylpropane because it is
simpler and more readily controlled than the widely-used preparation of
1-bromobutane.

We suggest you spend some time instructing your students on the assembly,
care and cleaning of ground-glass-joint apparatus, especially if this is the
first time they use it. They may also need help with the techniques of
simple distillation and the use of a separating funnel.

You may think it worth purchasing plastic clips (e.g. Griffin, QJM 200-W) to
hold joints firmly together and/or adjustable clamps with spring clips
(e.g. the Duoclamp from Spiring Enterprises, Horsham, Sussex) which enable
an assembly to be supported safely from a single retort stand and boss.

Students should not be discouraged by low yields at this stage. You should
discuss with them possible reasons for low yields.

Experiment 85.   Chemical properties of ethanol

To save time, you could allocate tests to different students and tell them
to share their results. Students working alone could omit G, and B or C,
and make use of the specimen results.

For Tollens' test in part B, a silver mirror will form only on a clean test-
tube. If the tube is greasy, a silver precipitate is formed instead.

For the Fehlings' test in part B, ethanal tends to form a resin in the
alkaline conditions, which may mask the formation of red copper(I) oxide.
The students may therefore note a greenish-yellow rather than red colour, which
is acceptable as a positive test for a reducing agent.

In test G (dehydration of ethanol) great care must be taken when heating the
ethanol on ceramic wool to prevent the pressure of the vapour causing the
pumice to block the tube. This experiment MUST be done in a fume cupboard
with the front down as far as is practically possible. Students may not
prepare enough ethene for both tests although, of course, each test may be
done on half a tube of gas. Ethene should not be tested with bromine
dissolved in an inert solvent because 1,2-dibromoethane is toxic.

Experiment 86. Chemical properties of phenol

The reaction between phenol and sodium can be violent and should be closely
supervised (or demonstrated, if you wish). Students should be provided with
very small pieces of sodium and not allowed to cut their own from stock.
Note that phenol is not sufficiently acidic to react with sodium
hydrogencarbonate.

Experiment 87. Reactions of amines

This experiment compares the reactions of phenylamine and butylamine with
those of ammonia. The reaction with hydrochloric acid gives a salt in each
case; phenylammonium chloride is immediately visible as a solid, whereas
butylammonium chloride and ammonium chloride are only visible after evapora-
tion of water.

The instructions for the reactions with nitrous acid are necessarily rather
complicated; we suggest that you make yourself familiar with the details so
that you can help students to follow them and understand what they are doing.
Note that we have preferred naphthalen-2-ol to naphthalen-1-ol because it
is less toxic.

Experiment 88. Observation and deduction exercise

This is intended primarily for students taking an A-level practical examination, in which one of the questions involves the identification of an unknown
compound. However, you may find it useful for assessment purposes.

Advise your students to study the wording of the question closely since the
type of answer required varies from paper to paper. For example, sometimes
details of additional tests must be given and sometimes equations and/or
comments on the type of reactions are required.

This test combines parts of two past practical examination papers, as
indicated below:

C = Glacial ethanoic acid, CH_3COOH
D = Ethanol, CH_3CH_2OH } London, Jan. 1978, Group I
E = Phenylammonium chloride, $C_6H_5NH_3{}^+Cl^-$ London, June 1979, Group I

Experiment 89. Reactions of aldehydes and ketones

In this experiment the students perform simple test-tube reactions on ethanal
and propanone. From their observations they should deduce that, although
there are similarities in the reactions of aldehydes and ketones, there are
some distinct differences.

In the reaction with sodium hydrogensulphite they will probably not obtain a
precipitate with ethanal; an addition product is formed but is so soluble
that it rarely crystallizes out.

The reaction with 2,4-dinitrophenylhydrazine may fail to give a precipitate if
too much aldehyde or ketone is used; the hydrazone may dissolve in excess
carbonyl compound.

In Fehling's test, ethanal tends to form a resin in the alkaline conditions,
which may mask the formation of red copper(I) oxide. The students may
therefore note a greenish-yellow (rather than a red) colour which is
acceptable as a positive test for a reducing agent.

In Tollen's test, a silver mirror will form only in a clean test-tube. Also,
it is important that only sufficient ammonia is added to barely dissolve the
silver oxide - some black specks of silver oxide should still remain.

Experiment 90. Identifying an unknown carbonyl compound

The student should be provided with an unknown aldehyde or ketone, selected
from the following list (preferably not the most familiar ones, methanal,
ethanal and propanone).

methanal (formaldehyde)	$HCHO$
ethanal (acetaldehyde)	CH_3CHO
propanal (propionaldehyde)	C_2H_5CHO
butanal (n-butyraldehyde)	C_3H_7CHO
2-methylpropanal (isobutyraldehyde)	$(CH_3)_2CHCHO$
benzaldehyde	C_6H_5CHO
propanone (acetone)	CH_3COCH_3
butanone (methyl ethyl ketone)	$CH_3COC_2H_5$
pentan-2-one (methyl n-propylketone)	$CH_3COC_3H_7$
pentan-3-one (diethyl ketone)	$C_2H_5COC_2H_5$
hexan-2-one (methyl n-butyl ketone	$CH_3COC_4H_9$
4-methylpentan-2-one (methyl isobutyl ketone)	$CH_3COCH_2CH(CH_3)_2$
cyclohexanone	$C_6H_{10}O$

In the first part of the experiment, the students identify the unknown
compound as either an aldehyde or a ketone. In the second part, they prepare
a derivative and determine its melting-point. The table of melting-points of
2,4-dinitrophenylhydrazones printed in the students' book should enable them
to identify the compound. They should check with you or your technician whether
they have identified X correctly.

Experiment 91. Chemical properties of carboxylic acids

In this experiment the students perform some simple test-tube reactions with glacial ethanoic acid. Aqueous ethanoic acid must <u>not</u> be used because the reactions of water with sodium and phosphorus pentachloride not only are violent but also mask the reactions of the —OH group in the acid, which are more moderate. You should closely supervise the use of sodium (or demonstrate if you wish) and not allow students to cut their own pieces from a large lump.

The triiodomethane (iodoform) reaction may fail to give a yellow precipitate if an iodine solution less concentrated than 10% is used.

From their observations, the students should conclude that the presence of an —OH group on the carbonyl carbon atom modifies the properties of both groups; carboxylic acids show virtually none of the properties of aldehydes and ketones and they are much more acidic than alcohols.

Experiment 92. Identifying salts of carboxylic acids

This experiment is intended primarily for students taking a practical examination which involves the identification of an unknown organic compound; if this is not a particular requirement of your syllabus, you may wish to omit it.

Students require access to practical textbook(s) which give details of the results of tests for cations, anions and organic functional groups.

I = calcium ethanoate (acetate), $(CH_3CO_2)_2Ca$
J = calcium methanoate (formate), $(HCO_2)_2Ca$ (London, Jan 1982)

Experiment 93. Chemical properties of ethanoyl chloride

In this experiment the students add ethanoyl chloride to water, ethanol, ammonia and phenylamine. Since some of these reactions are violent, especially those with ammonia and phenylamine, make sure your students follow the instructions carefully. They should perform the reactions at a fume cupboard with the safety glass pulled down as far as is practically possible. In view of the possible hazards in performing this experiment you may decide to carry it out yourself as a demonstration.

Experiment 94. Preparing an ester

This experiment gives students a second chance to perform the techniques of recrystallization and melting-point determination, which they first met in Experiment 90. If time is short and your particular syllabus does not specify that your students should know the details of the laboratory preparation of an ester, they could omit this experiment.

Experiment 95. Observation and deduction exercise

This is intended primarily for students taking an A-level practical examination
in which one of the questions involves the identification of an unknown
compound. We expect students to have access to practical textbook(s) and
notes.

The test is compiled from three examination questions but the answers are not
provided by the examining board.

F = 4-oxopentanoic acid (laevulinic acid), $CH_3COCH_2CH_2COOH$[*] (L1982)
G = benzaldehyde, C_6H_5CHO (L1982)
H = ethanoic anhydride (acetic anhydride), $(CH_3CO)_2O$ (L1978)

Note that the procedure for the triiodomethane reaction in test (d) is not
the one we have followed in our experiments. The test may fail if the sodium
chlorate(I) (hypochlorite) solution is not fresh; we suggest you check.

Old stock of ethanoic (acetic) anhydride may contain sufficient ethanoic acid
to react with phosphorus pentachloride. Again, we suggest you check.

In test (1) the ethanoic acid effectively dilutes the anhydride in order to
moderate the reaction.

Experiment 96. The glycine/copper(II) complex

This is a very simple experiment and should take only about 10 minutes.
We have assumed that students have dealt with complex ions already but, if
this is not the case, they will need some help in answering the questions.

Experiment 97. The biuret test for proteins

This experiment also is very short and straightforward but, as before, some
knowledge of complex formation is assumed. Make sure that students appreciate
the difference (in colour and in structure) between the complexes formed in
this experiment and the one in Experiment 96. The biuret test is given by
all proteins and all other compounds containing two or more peptide links.

Experiment 98. Paper chromatography

We have included this simple separation of amino acids for those students
who have had little or no experience of chromatography in their pre-A-level
course. It uses simple apparatus, small quantities of solvent and can be
completed in one hour.

Make sure that students are aware of the hazards of using ninhydrin as a
locating agent.

Experiment 99. Reactions of carbohydrates

This experiment is divided into five parts. If the recrystallization and melting-point determination is omitted, it can be completed in a $1\frac{1}{2}$-hour practical session. If your students do the polarimetry section, you may need to modify our instructions to suit your polarimeter.

The inversion of sucrose can be used as an experiment in reaction kinetics if you wish. In normal conditions, the concentrations of water and of hydrogen ions are constant, so that the reaction is first order with respect to sucrose. Since $(\alpha_t - \alpha_\infty) \propto [\text{sucrose(aq)}]$, students could measure the slope of their curve at a number of points to obtain a series of values of rate of reaction and $[\text{sucrose(aq)}]$, which give a straight line when plotted against one another.

The mono-derivatives with 2,4-dinitrophenylhydrazine are usually slow to form because they are fairly soluble in water - you may feel that this part of the experiment could be omitted. However, it is worth pointing out that the mon-derivatives of glucose and fructose, i.e. the products of condensation with the single $>C=O$ group in each case, are different whereas the osazones formed by heating with phenylhydrazine are identical. This is because the phenylhydrazine oxidizes the $>CHOH$ group at C^1 (fructose) or C^2 (glucose) to $>C=O$ so that condensation can occur at both C^1 and C^2. The osazone forms more rapidly for fructose because the terminal $-CH_2OH$ group is oxidized more readily than a $>CHOH$ group.

If a microscope is readily available, it might be interesting to show the osazone crystals and to point out that the appearance of osazones is a useful means of identification of sugars.

TECHNICIANS' SHEETS

for

Advanced Practical Chemistry

TECHNICIANS' SHEETS FOR ADVANCED PRACTICAL CHEMISTRY

The laboratory technician plays a vital role in the smooth running of any course which requires a lot of practical work. Because this book has been written in such a way as to encourage students to work at their own pace, several different experiments may be required at the same time. These technicians' sheets are designed to help you organise the materials required in such a way that experiments can be completed <u>successfully</u>, <u>safely</u> and <u>quickly</u>.

Requirements

Each experiment in the book has a technicians' sheet which lists the requirements for one student (or pair) in four sections - glassware, other apparatus, pure chemicals and solutions - in alphabetical order within each section. The teacher should tell you how many students (or pairs) are likely to do the experiment in each session, and the extent to which apparatus should be shared, so that you can easily work out the total requirements from the list.

<u>Trays</u>. We recommend that all the requirements for one experiment should be set out in a labelled tray (or trays) so that it can be taken in and re-issued easily over several practical sessions if necessary.

<u>Solutions</u>. We give instructions for making up all solutions. In general, the concentrations required are approximate, so that masses and volumes need not be measured with great accuracy - measuring cylinders and a balance of sensitivity ± 0.1 g are usually adequate. Where greater precision is required, we make it clear in the instructions. All aqueous solutions should, of course, be made up with distilled or deionized water.

<u>Quantities</u>. We list the minimum quantity of each chemical that each student (or pair) will need. There is usually no need to measure out this quantity; it simply enables you to check that you always supply sufficient for the students' needs.

<u>Chemical names</u>. We use systematic names (ASE) for chemicals throughout this book. However, since manufacturers' bottles are frequently labelled with traditional names, we give these as well (in brackets) where appropriate.

Dispensing chemicals

You may already have a system for dispensing chemicals of different types in convenient containers and safe quantities. If not, we suggest the following.

<u>Solids</u>. Either issue manufacturers' bottles (unless these are too large for convenience or safety) or dispense into small, <u>clearly labelled</u> screw-top jars.

<u>Liquids</u>. Organic liquids especially are best dispensed in manufacturers'
bottles (unless these are too large for convenience or safety). To help
students to handle these liquids safely, without spillage, and without
contaminating your stock, each bottle should be accompanied by a <u>labelled</u>
teat-pipette standing in its own small <u>labelled</u> beaker.

<u>Aqueous solutions</u>. Except when large quantities are required, e.g. for
titrations, solutions are best dispensed in small dropping bottles. If
standard reagent bottles are used, teat-pipettes should be issued together
with two small beakers to enable students to wash and rinse them in distilled
water before each use.

Safety

Take careful note of hazard symbols printed on manufacturers' labels. If
your stock bottles do not have modern labels, look at the requirements lists
in the students' books, where we print hazard symbols and some specific hazard
warnings. More details can be found in laboratory manuals (see below).

<u>Hazcards</u>. The School Science Service (formerly the CLEAPSE development
group) has produced a set of cards which not only state the potential dangers
of commonly used chemicals but also tell you what to do in the event of an
accident. It is a good idea to issue a HAZCARD with each dangerous chemical.
These cards can be purchased from the Association for Science Education or,
if your local authority subscribes, from the School Science Service, Brunel
University, Uxbridge, UB8 3PH.

Recommended books

These technicians' sheets are not intended to provide a complete guide to
laboratory management and safety. If you do not already have a reliable
modern manual, we suggest you consider·purchasing one or more of the following:

Creedy, J., 'A laboratory manual for schools and colleges'. (Heinemann)
ISBN 0 435 57130 3

Everett, K. and Jenkins, E., 'A safety handbook for science teachers'.
(Murray)
ISBN 0 7195 3760 6

SSSERC (Scottish Schools Science Equipment Research Centre), 'Hazardous
chemicals - a manual for schools and colleges'. (Oliver and Boyd)
ISBN 0 05 003204 6

Bretherick, L., 'Hazards in the chemical laboratory'. (Royal Society of
Chemistry)
ISBN 0 85186 419 8

Archenhold, W.F., Jenkins, E.W., Wood-Robinson, C., 'School Science
Laboratories - A handbook of design, management and organisation'. (Murray)
ISBN 0 7195 3436 4

ASE (The Association for Science Education), 'Chemical Nomenclature, Symbols
and Terminology'. (ASE, College Lane, Hatfield, Herts.)
ISBN 0 902786 53 9   (New edition in preparation for 1985.)

Experiment 1. Determining the Avogadro constant

Requirements per student (or pair)	Notes
measuring cylinder, 10 cm^3	
teat-pipette drawn out to a fine point or teat-pipette with adapter	- adapters called 'microtips' can be obtained from Hughes and Hughes Ltd. The teat-pipette should be capable of delivering about 100 drops to one cm^3.
large trough - glass or earthenware	
fine cotton thread, 40-50 cm	
scissors	
petroleum jelly or vaseline	
oleic acid solution in pentane (0.005%) (in corked test-tube)	- about 5 cm^3. 0.50 cm^3 of oleic acid made up to 50.0 cm^3 with pentane and then 0.50 cm^3 of this solution made up to 100 cm^3 with pentane for the final solution.
	If pentane is not available, petroleum ether (40-60, or even 30-40) can be used.

Experiment 2. Preparing a standard solution

Requirements per student (or pair)	Notes
beaker, 250 cm^3	
dropping pipette (teat-pipette)	
filter funnel	
stirring rod with rubber end	
volumetric flask, 250 cm^3, with stopper	
weighing-bottle	
spatula	
label for volumetric flask	
wash-bottle of distilled water	
safety spectacles	
access to balance capable of weighing to 0.01 g	
potassium hydrogenphthalate, $C_8H_5O_4K$, A.R.	- about 2 g. The students will be preparing a standard solution of potassium hydrogenphthalate which will be set aside in a labelled 250 cm^3 volumetric flask for Experiment 3.

Experiment 3. An acid-base titration

Requirements per student (or pair)	Notes
2 beakers, 100 cm^3	
burette, 50 cm^3	
4 conical flasks, 250 cm^3	
filter funnel, small	- to fill burette
pipette, 25 cm^3	
pipette filler	
burette stand	
safety spectacles	
wash-bottle of distilled water	
white tile	
standard potassium hydrogenphthalate solution	- prepared by the students and kept from Experiment 2
sodium hydroxide solution, $\sim$ 0.1 M NaOH	- at least 150 cm^3. See below.
phenolphthalein indicator solution	- see below.

Approx. 0.1 M sodium hydroxide solution. Dissolve 4.10 g of sodium hydroxide in freshly boiled and cooled distilled water. Make up to 1000 cm^3 and invert twenty times to ensure thorough mixing.

Phenolphthalein indicator solution. Dissolve 1 g of the reagent in 100 cm^3 of ethanol (IMS) and add 100 cm^3 of distilled water with constant stirring. Filter if a precipitate forms. Alternatively, the indicator solution can be bought already prepared for use.

Experiment 4. A redox titration

<u>Requirements per student (or pair)</u>	<u>Notes</u>
2 beakers, 100 cm^3
burette, 50 cm^3
4 conical flasks, 250 cm^3
filter funnel, small	- to fill burette
pipette, 10 cm^3
pipette filler
burette stand
safety spectacles
white tile
wash-bottle of distilled water
iodine solution, ~ 0.05 M I_2 (standardized)	- about 60 cm^3. See below.
sodium thiosulphate solution, 0.0500 M $Na_2S_2O_3$	- about 120 cm^3. See below.
starch indicator solution	- see below.

Since the accuracy of the students' results depends on the concentrations of the iodine and sodium thiosulphate solutions, these <u>MUST</u> be prepared as carefully as possible.

<u>Iodine solution 0.05 M I_2</u>. Dissolve 12.7 g of iodine crystals and 20 g of potassium iodide crystals in distilled water and make up to 1000 cm^3. This should be standardized against the sodium thiosulphate solution.

<u>Sodium thiosulphate solution, 0.0500 M $Na_2S_2O_3$</u>. Dissolve 12.41 g of A.R. sodium thiosulphate, $Na_2S_2O_3 \cdot 5H_2O$ in distilled water and make up to 1000 cm^3 If the solution is to be kept for more than a few days, add 0.10 g of sodium carbonate, Na_2CO_3, or 3 drops of trichloromethane (chloroform $CHCl_3$).

<u>Starch indicator solution</u>. Make a paste of 1 g of soluble starch with a little water, pour the paste, with constant stirring, into 100 cm^3 of boiling water, and boil for about 1 minute. Allow to cool and add about 2.0 g of potassium iodide. Keep the solution in a stoppered bottle. Test it before the experiment and renew it when necessary.

Experiment 5. A precipitation titration

Requirements per student (or pair)	Notes
2 beakers, 100 cm^3	
beaker, 250 cm^3	
burette, 50 cm^3	
4 conical flasks, 250 cm^3	
dropping pipette (teat-pipette)	
filter funnel, small	
pipette, 10 cm^3	
stirring rod with rubber end	
volumetric flask, 250 cm^3	
weighing-bottle	
burette stand	
label for flask	
pipette filler	
safety spectacles	
spatula	
wash-bottle of distilled water	
white tile	
access to balance capable of weighing to 0.01 g	
access to large "Silver Residues" bottle	- recovery of silver from the silver residues should be done every year or two, and silver nitrate regenerated. See also Experiment 45.
barium chloride crystals, $BaCl_2 \cdot 2H_2O$	- students must see the name only, NOT the formula.
sodium sulphate crystals	
silver nitrate solution, 0.0500 M $AgNO_3$	- about 150 cm^3. See below.
potassium chromate indicator solution	- 12.50 g of K_2CrO_4 made up to 250 cm^3 with distilled water.

<u>Silver nitrate solution 0.0500 M $AgNO_3$.</u> 8.49 g of A.R. $AgNO_3$ made up to 1000 cm^3 with distilled water and stored in an amber bottle.

If only commercial recrystallized silver nitrate (not A.R.) is available, the prepared solution must be standardized with sodium chloride solution, using potassium chromate indicator.

Experiment 6. A titration exercise

Requirements per student	Notes
burette, 50 cm^3	
3 conical flasks, 250 cm^3	
filter funnel, small	- to fill burette
pipette, 25 cm^3	
burette stand	
pipette filler	
safety spectacles	
wash-bottle of distilled water	
white tile	
dilute sulphuric acid ~ 1 M H_2SO_4	
solution A of ammonium iron(II) sulphate, $FeSO_4(NH_4)_2SO_4 \cdot 6H_2O$	- about 150 cm^3. See below.
solution B of potassium manganate(VII), $KMnO_4$, (potassium permanganate)	- about 150 cm^3. See below.

Solution A.   39.2 g of ammonium iron(II) sulphate, $FeSO_4 \cdot 6H_2O$ made up to 1000 cm^3 with distilled water and given the label: 'A-ammonium iron(II) sulphate, $FeSO_4(NH_4)_2SO_4 \cdot xH_2O$, 39.2 g dm^{-3}'

Solution B.   3.16 g of potassium manganate(VII) (permanganate) $KMnO_4$ made up to 1000 cm^3 with distilled water and then standardized using A.R. sodium ethandioate (oxalate) $Na_2C_2O_4$, to establish its precise concentration (close to 0.02 M).

This solution should be labelled as 'B-potassium manganate(VII), $KMnO_4$' with the precise concentration in mol dm^{-3} given to the students.

Experiment 7. Estimating the ionization energy of a noble gas

Requirements per student (or pair)

circuit board, with thyratron base and labelled sockets for leads

argon-filled triode, type 884

microammeter, 100 μA (0.1 mA) max.

voltmeter, 25 V, high resistance

protective resistor, ~ 200 Ω, 1 W

potential divider, 250 Ω, 3 W

power supplies:

   6.3 V (DC or AC) at about 3 A
   25 V DC at about 0.2 A (smoothed)
   3 V DC (2 x 1.5 V torch batteries)

12 connecting leads

Notes

- purchased boards* do not usually include a protective resistor, which is essential for this experiment. One can easily be made in school (see below for details).

- preferably mounted (see below).

- a rheostat can be used but, preferably, a wire-wound linear volume control** should be mounted on the circuit board (see below) so that students have only to connect to labelled, coloured sockets.

  * e.g. Griffin XLH-360-Y

  **e.g. RS Components 173/035

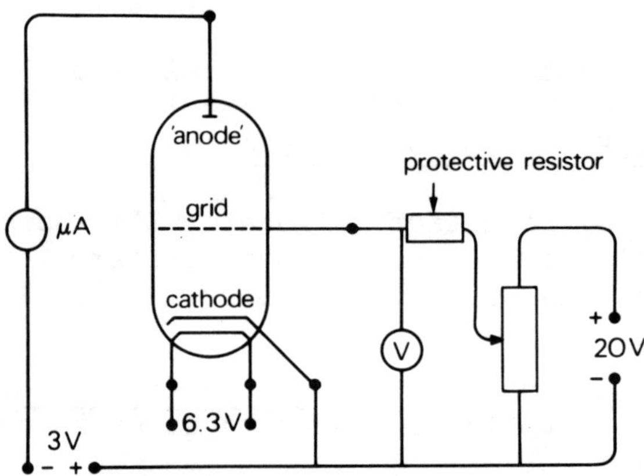

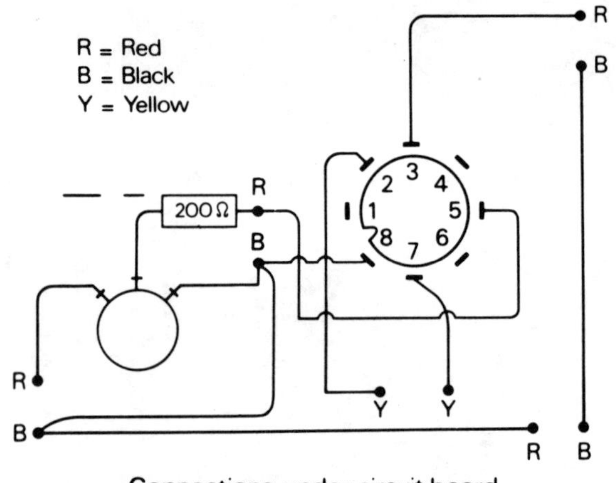

Connections under circuit board

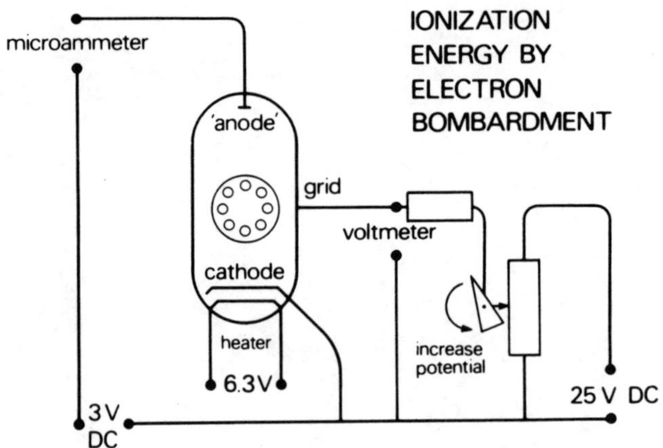

Markings on top of circuit board

Experiment 8. Using a hand spectroscope to observe the emission spectra of some S-block elements

Requirements per student (or pair)	Notes
beaker, 10 cm³	
2 watch-glasses	
Bunsen burner and bench mat	
flame-test wire	- preferably platinum
lamp, tungsten filament, pearl bulb	- room fixture is adequate
pestle and mortar, small	
safety spectacles	
spatula	
spectroscope, hand-held	
access to fume cupboard	
coloured plate of emission spectra	- consult the teacher
barium chloride, $BaCl_2$	
calcium chloride, $CaCl_2$	
lithium chloride, $LiCl$	- a few grams of each solid
sodium chloride, $NaCl$	
strontium chloride, $SrCl_2$	
hydrochloric acid, concentrated, HCl	

Experiment 9. Determining an enthalpy change of reaction

Requirements per student (or pair)	Notes
pipette, 25 cm³	
weighing-bottle	
balance (nearest 0.1 g)	
pipette filler	
polystyrene cup with lid	- expanded polystyrene to minimise heat loss, or two thin cups one inside the other. Lid needs hole for thermometer.
safety spectacles	
spatula	
stop-clock	- not needed if clock with second hand is visible in laboratory.
thermometer, 0-100°C	- preferably graduated in 0.2°C.
copper sulphate solution, 1.00 M $CuSO_4$	- about 30 cm³. Dissolve 250 g of $CuSO_4 \cdot 5H_2O$ in distilled water and make up to 1000 cm³.
zinc powder	- about 6 g. Granulated zinc is not suitable - reaction is too slow.

Experiment 10. Determining an enthalpy change of solution

Since this is a planning experiment, students will prepare their own lists of
apparatus. However, they are not likely to differ much from the list below.

Requirements per student (or pair)	Notes
weighing-bottle	
teat-pipette	- students may ask for a burette.
balance (nearest 0.01 g)	
polystyrene cup with lid	- <u>expanded</u> polystyrene to minimise heat transfer, or two thin cups one inside the other. Lid needs hole for thermometer.
safety spectacles	
spatula	
thermometer, 0-50°C	- must be graduated in 0.1°C.
ammonium chloride, NH_4Cl	- about 5 g.
distilled water	- about 100 cm^3.

Experiment 11. Using Hess's law

Requirements per student (or pair)	Notes
2 weighing-bottles	
teat-pipette	
balance (nearest 0.01 g)	
2 polystyrene cups with lids	- see Experiments 9 and 10
safety spectacles	
spatula	
thermometer, 0-50°C	- must be graduated in 0.1°C
magnesium sulphate (anhydrous), $MgSO_4$	- about 5 g. Heat to constant weight at about 250°C and store in desiccator. If necessary, the hydrated salt may be dehydrated by <u>prolonged</u> heating (with stirring) to constant weight.
magnesium sulphate-7-water, $MgSO_4 \cdot 7H_2O$	
distilled water	- about 100 cm^3.

Experiment 12. Another application of Hess's law

Requirements per student (or pair)	Notes
2 beakers, 100 cm³	
pipette, 50 cm³	– 25 cm³ can be used
2 weighing-bottles	
balance (to 0.01 g)	
Bunsen burner, tripod, gauze and bench protection mat	
safety spectacles	
spatula	
thermometer, 0–100°C	
vacuum flask with thermometer fitted (0–50°C)	– thermometer must be graduated in 0.1°C. See below for fitting.
anhydrous copper(II) sulphate, $CuSO_4$	– about 5 g, preferably in desiccator. See Experiment 14 for preparation.
copper(II) sulphate-5-water, $CuSO_4 \cdot 5H_2O$	– about 7 g. Grind finely with pestle and mortar for quick dissolving.
distilled water	– about 350 cm³. Tap water can be used for the first part only (250 cm³).

1. Insert a borer of size just larger than the thermometer into the hole in the bung from the narrow end.

2. Insert the thermometer to the required depth inside the borer.

3. Maintain the position of the thermometer while withdrawing the borer.

4. Remove the thermometer in a similar way after the experiment.

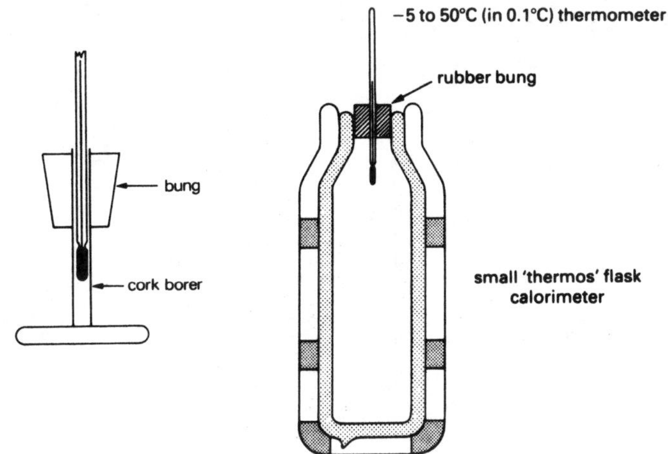

Experiment 13. Determining heats of combustion

Requirements per student (or pair)	Notes
6 beakers, 50 cm³	
6 teat-pipettes	- not required if alcohols are supplied in separate spirit lamps
heat of combustion apparatus	
Drechsel bottle	- to be fitted up as in diagram below. Consult teacher if alternative apparatus is used.
filter pump	
rubber tubing for connections	
small wood block	
spirit lamp(s), small	- 7 if possible, two filled with propanol and one for each of the other alcohols.
thermometer 0-50°C (in 0.1°C)	
balance (preferably to 0.001 g)	
Bunsen burner and protective mat	
labels	- for the spirit lamp(s)
retort stand with 2 clamps and bosses	- must be big enough to hold the heat of combustion apparatus.
safety spectacles	
tweezers	- not plastic-tipped - to adjust wick.
wood splints	
tap water in large bottle	- 2½ dm³. Stand in the laboratory overnight to get constant temperature.

Alcohols	Notes
propan-1-ol, C_3H_7OH	
butan-1-ol, C_4H_9OH	
pentan-1-ol, $C_5H_{11}OH$	
hexan-1-ol, $C_6H_{13}OH$	- a few cm³ of each. If possible they should be in labelled spirit lamps.
heptan-1-ol, $C_7H_{15}OH$	
octan-1-ol, $C_8H_{17}OH$	

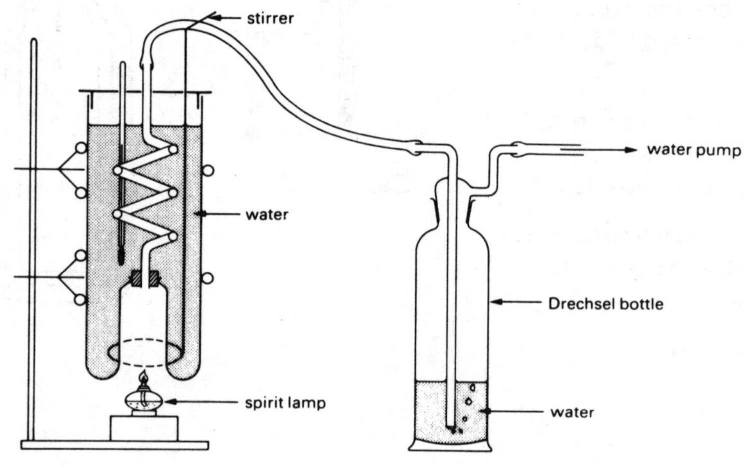

Experiment 14. A thermometric titration

Requirements per student (or pair)	Notes
burette, 50 cm³	
filter funnel, small	- to fill burette
pipette, 50 cm³	- 25 cm³ can be used
pipette filler	
polystyrene cup	
safety spectacles	
thermometer, 0-50°C	- graduated in 0.1°C
ethanoic (acetic) acid, ~ 2.0 M CH_3CO_2H	- about 50 cm³. 115 cm³ of glacial ethanoic acid made up to 1000 cm³
hydrochloric acid, ~ 2.0 M HCl	- about 50 cm³. 172 cm³ of concentrated HCl made up to 1000 cm³.
sodium hydroxide solution, 1.00 M NaOH	- about 100 cm³. If possible, make up precisely 1.00 M NaOH from a purchased ampoule. Otherwise dissolve 40.0 g of NaOH pellets in water, make up to 1000 cm³, and standardize against a known acid. All three solutions must be at the same temperature - make up the day before and store in the same room.

For Experiment 12

Anhydrous copper sulphate. It is not as easy as it sounds to dehydrate the blue crystals completely without overheating which causes a little dark copper oxide to appear. It is best to purchase the anhydrous salt each year but if this not practicable, proceed as follows:

1. Grind the required amount to powder, remembering that 25 g crystals reduces to 16 g. There is no point in dehydrating a lot more than will be used in a few days.

2. Put in a shallow dish (evaporating dish) of suitable size on a tripod and gauze. The powder should not be more than about 1 cm deep.

3. Heat over a moderate (not roaring) flame for at least 15 minutes, stirring constantly. Reduce heat if there is any sign of darkening in colour. Four molecules of water are lost very easily, but the fifth takes longer.

4. The only way to be certain that dehydration is complete is to heat to constant weight.

5. Allow to cool somewhat before storing in a desiccator.

Experiment 15. Making models of two metallic structures

Requirements per student (or pair)	Notes
31 expanded polystyrene spheres	- In a bag or box! Any size will do as long as they are all the same but large spheres are easier to handle.
Blu-tak or similar demountable adhesive	- Blu-tak is better for this purpose than Pritt 'Buddies'.

Experiment 16. Recognising ionic, covalent and metallic structures

Requirements per student (or pair)	Notes
beaker, 100 cm³	
6 ignition-tubes	
6 test-tubes	
test-tube holder	- to fit both tube sizes
test-tube rack	
battery, 6 V	
bulb, 6.5 V, 0.3 A	
bulb holder	
Bunsen burner and bench mat	
2 carbon electrodes	
3 connecting leads with crocodile clips	- only 2 needed if bulb holder is permanently mounted on battery
safety spectacles	
barium sulphate, $BaSO_4$	A
copper(II) oxide, CuO	B
graphite, C, powdered	C
glucose, $C_6H_{12}O_6$	D — a few grams of each in bottles or tubes labelled with letter only
potassium bromide, KBr	E
zinc, Zn, powdered	F
silica, SiO_2, precipitated (or fine white sand)	G

Experiment 17. Determining the molar mass of a gas

Requirements per student (or pair)	Notes
volumetric flask, 100 cm³, with stopper	- must be dry. A small plastic bottle may be used
thermometer, 0 °C to 100 °C	
carbon dioxide cylinder with regulator and needle valve	- if a CO_2 generator is used (e.g. Kipps) the gas should be bubbled through water and then concentrated H_2SO_4 to purify it.
rubber tubing, approx. 90 cm	- to connect generator to delivery tube
delivery tube, glass, approx. 20 cm	- must reach bottom of flask
balance, to weigh 100 g to 0.001 g	- 30 g maximum may be enough for a plastic bottle
balance, to weigh 200 g to 0.1 g	- must be able to weigh the flask full of water
barometer	- one reading per class per session, or telephone local meteorological office.

Experiment 18. Determining the molar mass of a volatile liquid

Requirements per student (or pair)	Notes
gas syringe, 100 cm³	- must be glass and have a free moving piston
steam generator with safety tube	
steam jacket for gas syringe	- see diagram and notes below
thermometer, 0 °C to 105 °C	- must fit steam jacket
hypodermic syringe, 2 cm³ and needle	- must be glass
silicone rubber, self-sealing	- to make temporary seal for needle; see, e.g. Griffin catalogue, or bathroom-sealant may be suitable
rubber cap, self-sealing	- to seal gas syringe nozzle
filter paper	- to wipe hypodermic needle
Bunsen burner and bench mat	
tripod and gauze	
balance, (to 0.001 g)	- shared
barometer	- one reading per class per session- or telephone local meteorological office
safety spectacles	
1,1,1-trichloroethane	- about 5 cm³

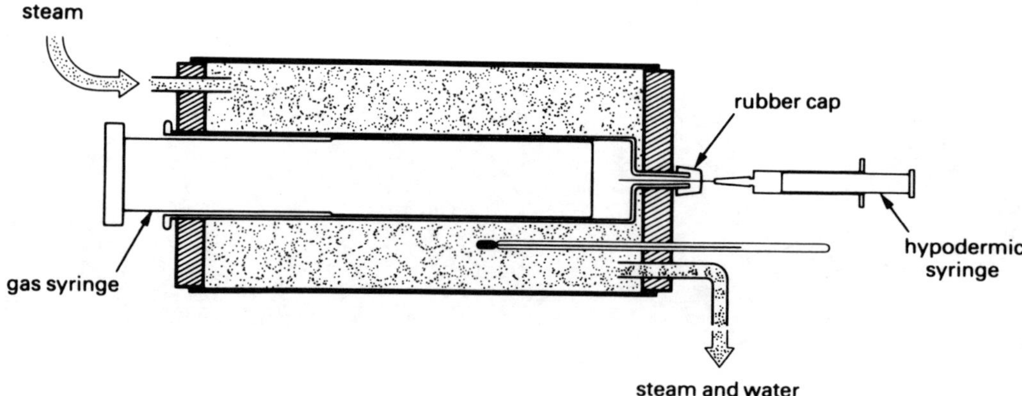

This apparatus may be purchased from laboratory suppliers or made in school. Plastic pipe such as is used in down-pipes from gutters can be used with wooden blocks sealing the ends. However, a window must be inserted to allow the volume of gas in the syringe to be measured. Alternatively, a polythene bag can be used with care, provided the arrangement is tested for safety - see J.Chem.Ed (March 1968) 45, 203.

Experiment 19. Determining the molar mass of a gas by effusion

<table>
<tr><td><u>Requirements per student (or pair)</u></td><td>Notes</td></tr>
</table>

Requirements per student (or pair)	Notes
gas syringe, 100 cm^3	- must be glass and have a free-moving piston
three-way tap	- lightly greased to be gas-tight
rubber connector, thick-walled	- to fit tap to syringe
aluminium foil, approx. 2 cm square	- cooking foil is ideal
quick-setting glue	- the teacher may ask you to glue the foil to the tap to seal one exit
retort stand, boss and clamp	
needle or sharp pin	
hydrogen cylinder, regulator and needle valve	- if a generator is used instead, the gas must be bubbled through water and then concentrated H_2SO_4 to purify it
2 lengths rubber tubing	- to connect 3-way tap to hydrogen cylinder and also to domestic gas tap
stopclock or stopwatch	- preferably to 0.1 sec

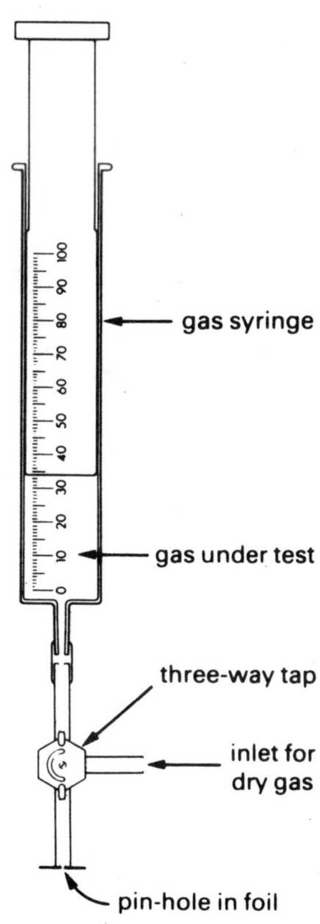

Experiment 20. The effect of concentration changes on equilibria

Requirements per student (or pair)	Notes
glass stirring rod	
4 test-tubes	– of equal size.
2 teat-pipettes	
safety spectacles	
spatula	
test-tube rack	
ammonium chloride, NH_4Cl	– about 2 g.
wash-bottle of distilled water	
iron(III) chloride solution, 0.5 M $FeCl_3$	– a few drops. Dissolve 13.5 g of hydrated solid in water, acidify with 12 cm^3 of concentrated hydrochloric acid and make up to 100 cm^3.
potassium thiocyanate solution, 0.5 M KCNS	– a few drops. Dissolve 5 g of solid in water and make up to 100 cm^3.

Experiment 21. Determining an equilibrium constant

Requirements per student (or pair)	Notes

Part A

2 measuring cylinders, 10 cm³	- one must be dry.
pipette, 5 cm³	
5 specimen tubes, 75 x 25 mm, with well fitting caps	
access to a balance	- sensitivity 0.01 g or better.
labels for tubes and stoppers	
safety pipette filler	
distilled water	- about 20 cm³.
ethyl ethanoate, $CH_3CO_2C_2H_5$	- about 20 cm³.
hydrochloric acid, 2 M HCl	- about 30 cm³. Add 172 cm³ of concentrated acid to water and make up to 1000 cm³.
safety spectacles	

Part B

burette, 50 cm³	
5 conical flasks, 250 cm³	
small funnel	- for burette filling
burette stand	
safety spectacles	
white tile	
wash-bottle of distilled water	
sodium hydroxide solution, 1 M NaOH (standardized)	- about 150 cm³. See below.
phenolphthalein indicator solution	- in a dropping bottle. Dissolve 0.1 g of solid in 60 cm³ of ethanol and make up to 100 cm³ with distilled water.

1 M NaOH, standardized. Weigh out 40 g of fresh pellets in a closed vessel, dissolve in freshly boiled, cooled, distilled water and make up to 1000 cm³ in a volumetric flask. Standardize by accurately diluting 25 cm³ to 250 cm³ and titrating with either standard HCl (methyl orange indicator) or potassium hydrogenphthalate (20.4 g dm⁻³) (phenolphthalein indicator). Write the concentration to three significant figures on the label. Alternatively ampoules, which can be diluted carefully to make precisely 1 M NaOH, are available from most chemical suppliers and are very convenient, if rather expensive.

Experiment 22. Determining a solubility product

Requirements per student (or pair)	Notes
4 reagent bottles, 250 cm³	– with stoppers
burette, 50 cm³	
4 conical flasks, 250 cm³	
4 filter funnels	– must be dry
small funnel	– for burette filling
measuring cylinder, 100 cm³	
pipette, 25 cm³	
filter papers	
labels for bottles	
burette stand	
safety spectacles	
safety pipette filler	
spatula	
thermometer, 0-100 °C (± 1 °C)	
white tile	
calcium hydroxide, $Ca(OH)_2$	– at least 10 g
distilled water	– at least 500 cm³
hydrochloric acid, 0.1 M HCl, (standardized)	– about 60 cm³. See below
phenolphthalein indicator solution	– in a dropping bottle

0.1 M HCl, standardized. Add 8.6 cm³ (or 10.3 g) of concentrated acid to distilled water and make up to 1000 cm³ in a volumetric flask. Standardize using 0.0500 M sodium carbonate solution (5.30 g of anhydrous Na_2CO_3 per dm³) and methyl orange indicator. Write the concentration to three significant figures on the label. Alternatively, ampoules, which can be diluted carefully to make precisely 0.1 M HCl, are available from most chemical suppliers and are very convenient, if rather expensive.

Experiment 23. Illustrating the common ion effect

Requirements per student (or pair)	Notes
teat-pipette	
2 test-tubes with corks	
safety spectacles	
spatula or forceps	
test-tube rack	
sodium hydroxide, NaOH	- a few pellets in a stoppered bottle.
hydrochloric acid, concentrated, HCl	- in a dropping bottle.
sodium chloride solution, saturated	- about 20 cm^3. Stir about 40 g of salt with 100 cm^3 of warm water and allow to cool. Leave excess solid in bottom of stock bottle.

Experiment 24. Distribution equilibrium

Requirements per student (or pair)	Notes
2 beakers, 100 cm³	
burette, 50 cm³, and stand	
2 conical flasks, 150 cm³	
measuring cylinder, 50 cm³	
pipette, 10 cm³	
separating funnel, 150 cm³	
safety spectacles	
safety pipette filler	
white tile	
wash-bottle of distilled water	
ammonia solution, ~ 1 M NH_3	- about 50 cm³. See below.
hydrochloric acid, 0.5 M HCl, (standardized)	- about 200 cm³. See below.
hydrochloric acid, 0.01 M HCl (standardized)	- about 200 cm³. See below.
1,1,1-trichloroethane, CH_3CCl_3	- about 50 cm³.
methyl orange indicator solution	- in a dropping bottle. See below.

<u>Approx. 1 M ammonia solution.</u> Add 56 cm³ of 0.880 ammonia to water, and make up to 1 dm³. Avoid breathing fumes, and any risk of ammonia getting into eyes.

<u>0.5 M HCl (standardized).</u> Add 43 cm³ of concentrated acid to water and make up to 1 dm³. Standardize with sodium carbonate solution, 0.250 M Na_2CO_3 (26.5 g of anhydrous sodium carbonate per dm³) and methyl orange indicator.

<u>0.01 M HCl (standardized).</u> Transfer, by pipette, 20.0 cm³ of 0.5 M HCl to a volumetric flask and make up to 1000 cm³. Alternatively, make the dilution approximately and standardize using 0.0100 M Na_2CO_3 (1.059 g of A.R. anhydrous salt per dm³) and methyl orange indicator.

<u>Methyl orange indicator solution.</u> Dissolve 0.04 g of dye in 20 cm³ of ethanol and make up to 100 cm³ with distilled water.

<u>Disposal of organic residues.</u> See Experiment 34.

Experiment 25. The pH of a weak acid at various concentrations

Requirements per student (or pair)	Notes
beaker, 50 cm³	
wash-bottle of distilled water	
pH meter and electrode	
ethanoic acid, 0.10 M CH_3CO_2H	– about 25 cm³. Dissolve 5.8 cm³ (or 6.0 g) of pure ethanoic acid (glacial acetic acid) in distilled water and make up to 1 dm³.
ethanoic acid, 0.010 M CH_3CO_2H	– about 25 cm³. Measure 1 volume of 0.10 M solution and make up to 10 volumes.
ethanoic acid, 0.0010 M CH_3CO_2H	– about 25 cm³. Measure 1 volume of 0.010 M solution and make up to 10 volumes.
ethanoic acid, 0.00010 M CH_3CO_2H	– about 25 cm³. Measure 1 volume of 0.0010 M solution and make up to 10 volumes.
buffer solution, pH 4.0	– about 25 cm³. Either follow the instructions supplied with commercial buffer tablets, or use the recipe below.

Buffer solution, pH 4.0

Dissolve 12.91 g of 2-hydroxypropane-1,2,3-tricarboxylic acid-1-water (citric acid), $C_6H_8O_7 \cdot H_2O$ and 10.95 g of anhydrous disodium hydrogen phosphate, Na_2HPO_4, in distilled water and make up to 1 dm³. If you have only the hydrated salt, $Na_2HPO_4 \cdot 12H_2O$, use 27.6 g.

Experiment 26. The pH of different acids at the same concentration

Requirements per student (or pair)	Notes

2 beakers, 50 cm^3

wash-bottle of distilled water

pH meter and electrode

benzoic acid solution,
  0.010 M $C_6H_5CO_2H$
 — about 25 cm^3. Dissolve 1.22 g of solid in about 800 cm^3 of hot water. Cool and make up to 1 dm^3.

boric acid solution,
  0.010 M H_3BO_3
 — about 25 cm^3. Dissolve 0.62 g of solid in water and make up to 1 dm^3.

dihydrogenphosphate(V) ion solution,
  0.010 M $H_2PO_4^-$
 — about 25 cm^3. Dissolve 1.56 g of solid sodium dihydrogenphosphate-2-water, $NaH_2PO_4 \cdot 2H_2O$ (or 1.36 g of KH_2PO_4) in water and make up to 1 dm^3.

ethanoic acid solution,
  0.010 M CH_3CO_2H
 — about 25 cm^3. Dissolve 0.58 cm^3 (or 0.60 g) of pure ethanoic acid (glacial acetic acid) in water and make up to 1 dm^3.

buffer solution, pH 4.0
 — about 25 cm^3. Either follow the instructions supplied with commercial buffer tablets, or use the recipe below.

Buffer solution, pH 4.0

Dissolve 12.91 g of 2-hydroxypropane-1,2,3-tricarboxylic acid-1-water (citric acid), $C_6H_8O_7 \cdot H_2O$ and 10.95 g of anhydrous disodium hydrogen phosphate, Na_2HPO_4, in distilled water and make up to 1 dm^3. If you have only the hydrated salt, $Na_2HPO_4 \cdot 12H_2O$, use 27.6 g.

Experiment 27. The action of a buffer solution

Requirements per student (or pair)	Notes
beaker, 50 cm^3	
2 beakers, 100 cm^3	
2 burettes	- these may be shared.
2 funnels, small	- for filling burettes.
measuring cylinder, 25 cm^3	
2 burette stands	
pH meter and electrode	- if not available, see below.
safety spectacles	
wash-bottle of distilled water	
pure water	- about 25 cm^3. This must be freshly deionized, or freshly boiled distilled water, and stoppered securely just before the lesson to prevent absorption of CO_2.
hydrochloric acid, 0.1 M HCl	- about 10 cm^3. Add 8.6 cm^3 of concentrated HCl to water and make up to 1 dm^3.
sodium hydroxide solution, 0.1 M NaOH	- about 10 cm^3. Dissolve 4.0 g of NaOH pellets in water and make up to 1 dm^3.
buffer solution, pH 7	- about 25 cm^3. Either follow the instructions supplied with commercial tablets or use the recipe below.

If no pH meter is available

thin glass stirring rod
full range pH paper, pH 1-14
narrow range pH papers to
  cover pH 2-12

Buffer solution, pH 7

Dissolve <u>one</u> of the dihydrogenphosphates from list A and <u>one</u> of the
hydrogenphosphates from list B in distilled water and make up to 1 dm^3.

A		B	
$NaH_2PO_4 \cdot 2H_2O$	- 6.08 g	$Na_2HPO_4 \cdot 12H_2O$	- 21.85 g
KH_2PO_4	- 5.31 g	Na_2HPO_4	- 8.66 g
		K_2HPO_4	- 10.62 g

Experiment 28. Preparation of buffers: testing their buffering capacity and the effect of dilution

Since this experiment is designed as a planning exercise, students will probably be asked to submit their own list of requirements, but the following list should be adequate. The experiment is similar to Experiment 27.

<u>Requirements per student (or pair)</u>

<u>Notes</u>

2 beakers, 100 cm^3	
beaker, 50 cm^3	
2 burettes, 50 cm^3	
2 funnels, small	– to fill burettes.
measuring cylinder, 25 cm^3	
stirring rod, thin glass	
2 burette stands	
pH meter	
pH paper, full range	– required only if pH meter is not available.
pH papers, narrow range	
safety spectacles	
wash-bottle of distilled water	
ammonia solution, 1.0 M NH$_3$	– about 30 cm^3. Carefully add 56 cm^3 of 0.880 NH$_3$ to water and make up to 1 dm^3. This solution should preferably be standardized. Alternatively, make up from commercial ampoule by accurate dilution.
ammonium chloride solution, 1.0 M NH$_4$Cl	– about 80 cm^3. Dissolve 53.5 g solid in water and make up to 1 dm^3.
ethanoic (acetic) acid, 1.0 M CH$_3$CO$_2$H	– about 30 cm^3. Add 58 cm^3 of glacial acid to water and make up to 1 dm^3. This solution should preferably be standardized. Alternatively, make up from commercial ampoules by accurate dilution.
sodium ethanoate (acetate) solution, 1.0 M CH$_3$CO$_2$Na	– about 80 cm^3. Dissolve 82.03 g of the anhydrous salt or 136.1 g of the trihydrate in water and make up to 1 dm^3.

Experiment 29. Determining the pH range of some acid-base indicators

Requirements per student (or pair)	Notes
3 beakers, 100 cm^3	
4 beakers, 250 cm^3	
3 burettes	- two may be shared.
3 funnels, small	- to fill burettes,
pipette, 25 cm^3	
3 burette stands	
pH meter and electrode	- if no meter is available, students can calculate the pH.
pipette filler	
safety spectacles	
wash-bottle of distilled water	
hydrochloric acid, 0.1 M HCl	- about 125 cm^3. Add 8.6 cm^3 of concentrated HCl to water and make up to 1 dm^3.
sodium hydroxide solution, 0.10 M NaOH	- about 125 cm^3. Dissolve 4.00 g of NaOH pellets in water and make up to 1 dm^3. If this is done carefully, the solution need not be standardized.
buffer solution, pH 3.1 (or 3.25)	- about 125 cm^3. Follow the instructions supplied with BDH Universal Buffer Mixture, which has an unpleasant smell, or use the recipe below.
litmus indicator	- dissolve 1.0 g of solid in boiling water, filter and make up to 100 cm^3.
bromophenol blue indicator	- Dissolve in ethanol. 0.04 g + 20 cm^3 C$_2$H$_5$OH
methyl orange indicator	- 0.04 g + 20 cm^3 C$_2$H$_5$OH
methyl red indicator	- 0.02 g + 20 cm^3 C$_2$H$_5$OH
phenolphthalein indicator	- 0.10 g + 60 cm^3 C$_2$H$_5$OH

(The four indicators above: Make up to 100 cm^3 with H$_2$O)

<u>Universal buffer solution, pH 3.25</u>

Dissolve the following materials in distilled water and make up to 1 dm^3:

boric acid, H$_3$BO$_3$	2.47 g
potassium dihydrogenphosphate, KH$_2$PO$_4$	5.44 g
(<u>or</u> sodium dihydrogenphosphate-2-water, NaH$_2$PO$_4 \cdot$2H$_2$O)	(6.24 g)
ethanoic acid, CH$_3$CO$_2$H, (glacial acetic acid)	2.40 g
(<u>or</u> ethanoic acid solution, 1.00 M CH$_3$CO$_2$H)	(40.0 cm^3)

Experiment 30. Determining the ionization constant for an indicator

Requirements per student (or pair)	Notes
Method 1	
measuring cylinder, 10 cm³	
stirring rod	
2 teat-pipettes	
20 test-tubes	– all the same size.
test-tube rack or racks	– to hold 2 rows of 9 tubes.
safety spectacles	
wash-bottle of distilled water	
bromophenol blue solution	– about 10 cm³. Dissolve 0.04 g of solid in 20 cm³ of ethanol and make up to 100 cm³ with distilled water.
buffer solution, pH 3.7	– about 10 cm³. See below.
hydrochloric acid, concentrated, HCl	– a few drops.
sodium hydroxide solution, 4 M NaOH	– a few drops. Carefully dissolve 16 g of NaOH pellets in water and make up to 100 cm³.
Method 2	
beaker, 100 cm³	
2 beakers, 250 cm³	
measuring cylinder, 25 cm³	
stirring rod	
Bjerrum wedge	– this can be made from perspex. See Experiment 31.
safety spectacles	
bromophenol blue solution	– about 20 cm³. See Method 1.
buffer solution, pH 3.7	– about 25 cm³. See below.
hydrochloric acid, 0.1 M HCl	– about 250 cm³. Add 8.6 cm³ of concentrated acid to water and make up to 1 dm³.
sodium hydroxide solution, 0.1 M NaOH	– about 250 cm³. Dissolve 4.0 g of NaOH pellets in water and make up to 1 dm³.

Buffer solution, pH 3.7

Dissolve 24.31 g of disodium hydrogenphosphate(V)-12-water, $Na_2HPO_4 \cdot 12H_2O$, (or 9.62 g of anhydrous Na_2HPO_4, or 11.79 g of K_2HPO_4) and 6.94 g of 2-hydroxypropanone-1,2,3-tricarboxylic acid-1-water (citric acid), $C_3H_4OH(CO_2H)_3 \cdot H_2O$ in water and make up to 1 dm³.

Experiment 31. Determining the dissociation constant
of a weak acid using an indicator

<u>Requirements per student (or pair)</u> | Notes

<u>Method 1</u>

measuring cylinder, 10 cm³

stirring rod

2 teat-pipettes

20 test-tubes
- all the same size.

test-tube rack or racks
- to hold 2 rows of 9 tubes.

safety spectacles

wash-bottle of distilled water

bromophenol blue solution
- about 10 cm³. Dissolve 0.04 g of solid in 20 cm³ of ethanol and make up to 100 cm³ with distilled water.

benzoic acid solution
  0.02 M $C_6H_5CO_2H$
- about 5 cm³. Dissolve 2.44 g of solid in 800 cm³ of hot water, cool and make up to 1 dm³.

sodium benzoate solution,
  0.02 M $C_6H_5CO_2Na$
- about 5 cm³. Dissolve 2.88 g of anhydrous solid in water and make up to 1 dm³.

hydrochloric acid, concentrated, HCl
- a few drops.

sodium hydroxide solution, 4 M NaOH
- a few drops. Carefully dissolve 16 g of NaOH pellets in water and make up to 100 cm³.

For Method 2, see next page.

Experiment 31. Determining the dissociation constant
of a weak acid using an indicator

<u>Requirements per student (or pair)</u>   <u>Notes</u>

<u>Method 2</u>

beaker, 100 cm^3

2 beakers, 250 cm^3

measuring cylinder, 25 cm^3

stirring rod

Bjerrum wedge   - this can be made from perspex.
                                                     See below.

safety spectacles

bromophenol blue solution   - about 20 cm^3. Dissolve
                                                     0.04 g of solid in 20 cm^3 of
                                                     ethanol and make up to
                                                     100 cm^3 with water.

benzoic acid solution,   - about 15 cm^3. Dissolve
   0.02 M $C_6H_5CO_2H$ 2.44 g of solid in 800 cm^3
                                                     of hot water, cool and make
                                                     up to 1 dm^3.

sodium benzoate solution,   - about 15 cm^3. Dissolve
   0.02 M $C_6H_5CO_2Na$ 2.88 g of anhydrous solid
                                                     in water and make up to
                                                     1 dm^3.

hydrochloric acid, 0.1 M HCl   - about 250 cm^3. Add 8.6 cm^3
                                                     of concentrated acid to water
                                                     and make up to 1 dm^3.

sodium hydroxide solution,   - about 250 cm^3. Dissolve
   0.1 M NaOH   4.0 g of NaOH pellets in
                                                     water and make up to 1 dm^3.

Making a Bjerrum wedge

To get a general idea of the construction, look at a laboratory supplier's
catalogue and the diagrams on pages 71 and 72 of the book. We suggest that you
cut pieces of perspex, or other transparent plastic, and glue them together
using a suitable commercial adhesive or a solvent such as dichloromethane.
Clean, quick-setting and leak-proof joints can be made by touching one edge
onto the surface of a solvent for a few moments and then pressing quickly into
position, but we suggest you perfect a gluing technique using scraps before
starting the construction.

Cut a base, measuring 20.0 cm x 5.5 cm, two side pieces, 20.0 cm x 5.0 cm, and
a diagonal divider, 20.5 cm x 5.0 cm. Glue the side pieces and the diagonal
down on to the base and, when the glue is set, support the wedge in a vice and
file both ends, where necessary, so that end pieces can be fitted perfectly
flush. Cut end pieces to size and glue carefully in position. Fill each side
of the wedge separately to check for leaks.

Make the upper cell 5 cm x 5 cm x 2 cm wide, but make the ends 5 cm wide and
10 cm deep so that the cell slides along the wedge. Attach masking strips so
that, when comparing colours, only a narrow strip of the wedge can be seen
immediately below the upper cell.

Experiment 32. Obtaining pH curves for acid-alkali titrations

Requirements per student (or pair)	Notes
4 beakers, 250 cm³	
2 burettes, 50 cm³	- some sharing may be possible.
2 funnels, small	
beaker, 100 cm³	
pipette, 25 cm³	
2 burette stands	
4 labels for beakers	
pH meter and electrode	- if not available, see below
pipette filler	
safety spectacles	
stirrer, magnetic or electric	
wash-bottle of distilled water	
ammonia solution, 0.100 M HCl	- about 35 cm³. See below.
ethanoic (acetic) acid, 0.100 M CH₃CO₂H	- about 25 cm³. See below.
hydrochloric acid, 0.100 M HCl	- about 25 cm³. See below.
sodium hydroxide solution, 0.100 M NaOH	- about 35 cm³. See below.
buffer solution, pH 7	- to calibrate pH meter. See Experiment 27.

If no pH meter is available

thin glass stirring rod

pH paper, pH 1-14

narrow range pH papers to
   cover pH 1-14

0.100 M solutions of acids and alkalis

The concentrations of these solutions must be precisely known <u>or</u> precisely
equal, and preferably both. Ampoules of concentrated solutions, which can be
diluted to give precisely 0.100 M solutions, are available from most
chemical suppliers. Although the ampoules are quite expensive, they may be
worth buying for special purposes such as this.

To equalise the concentrations of approximately 0.1 M acids and alkalis

Titrate 25.0 cm³ of A against B in the usual way. Let the titre be V cm³.

If $V > 25$, add $40 \times (V-25)$ cm³ of water to 1000 cm³ of A.

If $V < 25$, add $1000 \times \left(\dfrac{25-V}{V}\right)$ cm³ of water to 1000 cm³ of B.

Experiment 33. The variation of boiling-point with composition for a binary liquid mixture

Requirements per student (or pair)	Notes

Requirements per student (or pair)

2 burettes, 50 cm^3

2 funnels, small

ground-glass-joint apparatus

test-tube

thermometer, 0-100 °C ± 0.2 °C

anti-bumping granules

Bunsen burner, gauze, tripod and mat

safety spectacles

cyclohexane, C_6H_{12}

ethanol, C_2H_5OH

methyl ethanoate, $CH_3CO_2CH_3$

propan-1-ol, $CH_3CH_2CH_2OH$

propan-2-ol, $CH_3CH(OH)CH_3$

trichloromethane, $CHCl_3$

Notes

- must be dry.

- for filling burettes and flasks

- for <u>one</u> of the sets shown below.

- a long stem thermometer, graduated in 1°, will serve if necessary. A 50-100 °C thermometer, graduated in 0.1° would be ideal <u>provided</u> the scale is fully visible.

- ~ 20 cm^3 of each liquid. Each student will probably need only only <u>two</u> of these - consult the teacher. IMS (74 op) will serve as ethanol.

<u>Disposal of organic solvents.</u> See Experiment 34.

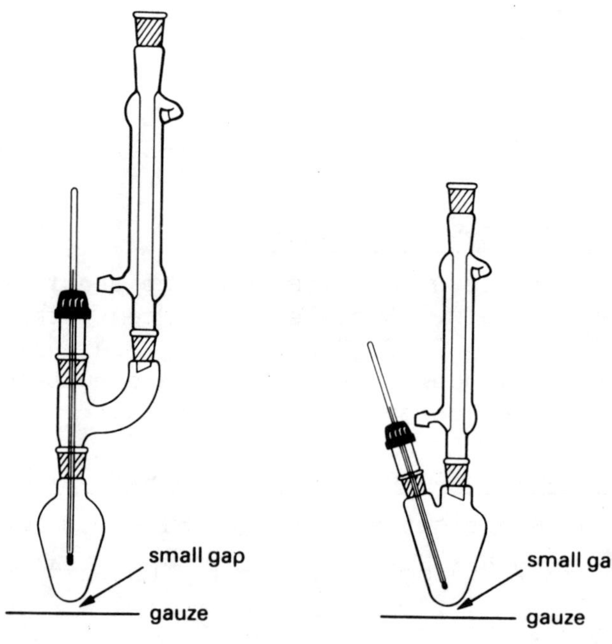

Experiment 34. Measuring temperature changes on forming solutions

Requirements per student (or pair)	Notes
beaker, 250 cm^3	- preferably tall form, but not essential.
boiling-tube	
2 funnels, small	- for transferring liquid.
2 measuring cylinders, 10 cm^3	
thermometer, -5 to 50 $^{\circ}$C, $\pm$ 0.1 $^{\circ}$C	
cotton wool	- or similar insulating material, sufficient to fill beaker
safety spectacles	
cyclohexane, C_6H_{12}	- ~ 10 cm^3 of each. IMS (74 op) is suitable for ethanol. The experiment is simplified if all are kept at the same temperature for 24 hours beforehand.
ethanol, C_2H_5OH	
methyl ethanoate, $CH_3CO_2CH_3$	
propan-1-ol, $CH_3CH_2CH_2OH$	
propan-2-ol, $CH_3CH(OH)CH_3$	
trichloroethane, $CHCl_3$	

Disposal of organic residues

Only the <u>smallest</u> quantities may be poured down a sink in a fume cupboard
with plenty of flowing water. It is better to collect residues and dispose
of them quickly by burning in a metal tray in the open away from people and
buildings. Alternatively, they can be allowed to evaporate from a metal
tray, covered with a mesh, on a roof with controlled access. Normally,
residues should not be stored, nor should residues from different experiments,
containing different chemicals, be mixed. For disposal of larger quantities,
consult your local authority.

Experiment 35. Determining the approximate strength of a hydrogen bond

Students will be asked to make their own lists. Their requirements will
probably be similar to those for Experiment 34 except that only two liquids
are required, trichloromethane and methyl ethanoate. If you are in doubt
about the suitability of students' requests, consult the teacher.

Experiment 36. The effect of hydrogen bonding on liquid flow

 | Notes

stopclock or stopwatch

4 long glass tubes, sealed at one end and securely corked at the other, each containing one of the following liquids:

propan-1-ol, $CH_3CH_2CH_2OH$

propan-1,2-diol, $CH_3CH(OH)CH_2OH$

propan-1,2,3-triol (glycerol), $CH_2(OH)CH(OH)CH_2OH$

propane-1-2,3-triyl triethanoate (glycerol triacetate), $CH_2(OCOCH_3)CH(OCOCH_3)CH_2OCOCH_3$

Notes

- tubes must be labelled and of equal length (~ 1 m) and diameter (~ 8 mm). Each must have an air bubble of the same size (~ 20 mm). One set will serve a large class, and the tubes may be stored for other occasions. Corks must fit well enough to allow the tubes to be inverted.

Experiment 37. Testing liquids for polarity

Requirements per student (or pair) | Notes

9 beakers, 250 cm^3

9 burettes in stands, each corked and labelled and containing one of the following liquids:

cyclohexane, C_6H_{12}

cyclohexene, C_6H_{10}

distilled water, H_2O

hexane, C_6H_{14}

hexan-1-ol, $C_6H_{13}OH$

methyl ethanoate, $CH_3CO_2CH_3$

propanone, CH_3COCH_3

tetrachloromethane, CCl_4

trichloromethane, $CHCl_3$

9 small funnels

rod of polythene or perspex or ebonite

piece of fur or suitable cloth

Notes

- must be dry

- set up in the fume cupboard. Burettes should all be of the same type so that they give the same rate of flow and the same volume of liquid (at least 25 cm^3) placed in each. The experiment may still be done if some of the liquids are not available. One set will serve a large class.

- for refilling burettes.

- must be quite dry.

- must be quite dry. It is worth experimenting to find a good combination so that the rod is readily charged when rubbed with the cloth.

Experiment 38. Determining relative molecular mass by a freezing-point method

Requirements per student (or pair)	Notes
2 beakers, 400 cm^3	
2 boiling-tubes, with stoppers	– must be dry.
thermometer, 0-100 °C	– long stem.
Bunsen burner, tripod, gauze and mat	
retort stand, boss and clamp	
access to balance (sensitivity 0.01 g)	
safety spectacles	
spatula	
stopclock	
1,4-dichlorobenzene, $C_6H_4Cl_2$, labelled compound A	– ~ 40 g.
naphthalene, $C_{10}H_8$, labelled compound B	– ~ 3 g.

Experiment 39. Determining enthalpy changes and volume changes of solution

Requirements per student (or pair)	Notes
2 burettes with stoppers to fit thermometer, 0-100 °C (± 0.5 °C)	- a gas-measuring tube or gas burette, if available, might be preferable - consult the teacher.
balance, sensitivity 0.01 g	
expanded polystyrene cup	- for use as a calorimeter
protective gloves (disposable)	- not required if $FeCl_3$ is unavailable.
2 retort stands, bosses and clamps	
safety spectacles	
spatula	
wash-bottle of distilled water	
weighing bottle with tight-fitting cap or stopper	
calcium chloride, anhydrous, $CaCl_2$	- ~ 25 g. Heat granules gently to constant weight. Grind to a powder and store in a desiccator.
iron(III) chloride, anhydrous, $FeCl_3$	- 35 g. Anhydrous $FeCl_3$ can be purchased but is very hygroscopic and does not store well. Keep in a desiccator.
lithium chloride, LiCl	- ~ 10 g.
potassium chloride, KCl	- ~ 15 g. Dry in a warm oven, grind to a powder and store in a desiccator.
sodium chloride, NaCl	- ~ 15 g.

Experiment 40. Investigating the hydrolysis of benzenediazonium chloride

Requirements per student (or pair)	Notes
beaker, 250 cm^3	
glass syringe, 100 cm^3	- not greased. Piston must move freely. Diagram below.
graduated pipette, 5 cm^3	- 10 cm^3 will serve if necessary.
measuring cylinder, 10 cm^3	
side-arm test-tube, with bung	
teat-pipette	
test-tube	
thermometer, 0-100 °C	
three-way tap	- greased lightly. Must be leakproof.
balance, sensitivity ± 0.01 g	
crushed ice	- ~ 200 cm^3. Enough to cool a test-tube.
pipette filler	
protective gloves	
pumice or anti-bumping granules	
2 retort stands, clamps and bosses	
rubber tubing, thick wall	- two short pieces to connect 3-way tap to syringe and side-arm tube.
safety spectacles	
spatula	
stopclock or stopwatch	
wash-bottle of distilled water	
water-bath, thermostatically controlled	- if possible, set this up before the lesson and allow to warm up to 40-50 °C.
hydrochloric acid, concentrated, HCl	- ~ 3 cm^3.
phenylamine, $C_6H_5NH_2$ (aniline)	- ~ 1 cm^3.
sodium nitrite, $NaNO_2$	- ~ 1 g.

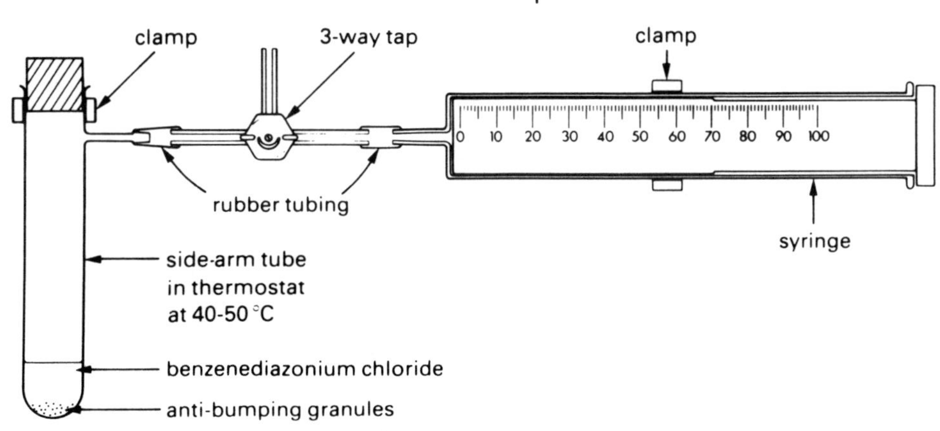

Experiment 41. The kinetics of the reaction between iodine and propanone in aqueous solution

<table>
<tr><td>

<u>Requirements per student (or pair)</u>

4 beakers, 100 cm^3

4 burettes, 50 cm^3, with stands

4 funnels, small

6 test-tubes

thermometer, 0-100 °C

</td><td>

<u>Notes</u>

</td></tr>
</table>

Requirements per student (or pair)	Notes
4 beakers, 100 cm^3	
4 burettes, 50 cm^3, with stands	- may be shared by several students.
4 funnels, small	
6 test-tubes	- preferably optically matched, as supplied for use with colorimeters. Otherwise, select tubes which look identical and fit the colorimeter.
thermometer, 0-100 °C	
6 bungs or corks	- to fit the test-tubes.
colorimeter, with set of filters	
safety spectacles	
stopclock or stopwatch	
wash-bottle of distilled water	
iodine solution, 0.020 M I$_2$, in KI(aq)	- ~ 30 cm^3. Dissolve 5.08 g of iodine and 16 g of KI in distilled water and make up to 1000 cm^3. Iodine dissolves very slowly - frequent shaking or stirring is necessary to ensure that it dissolves <u>completely</u>.
hydrochloric acid, 2.0 M HCl	- ~ 20 cm^3. Add 172 cm^3 of concentrated acid to water and make up to 1000 cm^3.
propanone solution, 2.0 M CH$_3$COCH$_3$ (acetone)	- ~ 20 cm^3. Add 147 cm^3 of pure propanone to water and make up to 1000 cm^3.

Experiment 42. Determining the activation energy of a reaction

Requirements per student (or pair)	Notes
beaker, 400 cm^3	- to be used as a water-bath.
4 beakers, 100 cm^3	
4 burettes, 50 cm^3, with stands	- may be shared by several students.
4 funnels, small	
2 boiling-tubes	
2 thermometers, 0-100 °C	
Bunsen burner, tripod, gauze and mat	
retort stand, clamp and boss	
safety spectacles	
stopclock or stopwatch	
potassium iodide solution, 0.50 M KI	- ~ 25 cm^3. Dissolve 83 g of KI in water and make up to 1000 cm^3.
potassium peroxodisulphate(VI) (persulphate) solution, 0.020 M $K_2S_2O_8$	- ~ 50 cm^3. Dissolve 5.4 g of $K_2S_2O_8$ (or 4.8 g of $Na_2S_2O_8$) in water and make up to 1000 cm^3.
sodium thiosulphate solution, 0.010 M $Na_2S_2O_3$	- ~ 25 cm^3. Dissolve 2.48 g of $Na_2S_2O_3 \cdot 5H_2O$ in water and make up to 1000 cm^3.
starch solution, 0.2%	- ~ 15 cm^3. Mix 0.2 g of 'soluble' starch into a paste with cold water and pour, with constant stirring, into 100 cm^3 of boiling water.

Experiment 43. Determining the activation energy of a catalysed reaction

This experiment is a planning exercise for the students, who should work out
their own list of requirements and hand it in to you, via the teacher. They
will probably request the same apparatus and chemicals as are listed above,
except that the concentration of one (or more) of the solutions may be
smaller - preferably the potassium iodide. If in doubt about the suitability
of students' requests, consult the teacher. In addition to the above,
students will need a few cm^3 of an iron(III) solution. We suggest the
following:

iron(III) chloride solution, 0.05 M $FeCl_3$	- ~ 5 cm^3. Dissolve 1.4 g of hydrated solid in water with 5 cm^3 of concentrated hydrochloric acid and make up to 100 cm^3.

Experiment 44. A bromine 'clock' reaction

<u>Requirements per student (or pair)</u> | Notes

2 beakers, 100 cm^3

3 beakers, 100 cm^3

3 burettes, 50 cm^3, and stands

3 funnels, small

- may be shared by
  several students

measuring cylinder, 25 cm^3

thermometer, 0-100 °C

safety spectacles

stopclock or stopwatch

wash-bottle of distilled water

white tile

methyl orange solution, 0.001%, acidified,
  labelled SOLUTION C

- ~ 200 cm^3. Add 20 cm^3 of methyl
  orange solution (containing
  0.5 g dm^{-3}) to 400 cm^3 of
  1 M H_2SO_4 and make up to
  1000 cm^3.

methyl orange solution, 0.001%, in
  0.40 M KBr, labelled SOLUTION D

- ~ 100 cm^3. Dissolve 48 g of
  KBr in water, add 20 cm^3 of
  methyl orange solution (containing 0.5 g dm^{-3}) and make
  up to 1000 cm^3.

phenol solution, 0.0001 M C_6H_5OH

- ~ 100 cm^3. Make a 0.010 M
  solution by dissolving 1.0 g
  of phenol in 1000 cm^3 of water.
  Dilute 10 cm^3 of this to
  1000 cm^3 immediately before use.

potassium bromate(V) solution,
  0.20 M $KBrO_3$

- ~ 100 cm^3. Dissolve 33.4 g of
  $KBrO_3$ in water and make up to
  1000 cm^3.

potassium bromate(V) solution,
  0.0050 M $KBrO_3$

- ~ 120 cm^3. Take 25.0 cm^3 of
  0.20 M $KBrO_3$ and make up to
  1000 cm^3 with distilled water.

potassium bromide solution, 0.010 M KBr

- ~ 120 cm^3. Dissolve 1.2 g of
  KBr in water and make up to
  1000 cm^3.

sulphuric acid, 0.010 M H_2SO_4

- ~ 50 cm^3. Dilute 10 cm^3 of
  1.0 M H_2SO_4 to 1000 cm^3. To
  make 1.0 M H_2SO_4, add 55 cm^3
  of concentrated acid, slowly
  and with constant stirring, to
  800 cm^3 of water and make up
  to 1000 cm^3.

Note that, while these solutions need not be made up with great accuracy, it
is essential that an individual student has sufficient from the <u>same stock</u> to
complete each part of the experiment.

Experiment 45. Some simple redox reactions

Requirements per student (or pair)	Notes
6 test-tubes	
emery paper	- for cleaning metal strips.
safety spectacles	
test-tube rack	
copper, foil, Cu	- two strips, 5 cm x 0.5 cm.
silver, wire, Ag	- two lengths, 5 cm.
zinc, foil, Zn	- two strips, 5 cm x 0.5 cm.
copper(II) sulphate solution, 0.50 M $CuSO_4$	- ~ 10 cm^3. Dissolve 12.5 g of $CuSO_4 \cdot 5H_2O$ in water and make up to 100 cm^3.
silver nitrate solution, 0.10 M $AgNO_3$	- ~ 10 cm^3. Dissolve 1.7 g of solid in water and make up to 100 cm^3. Store and dispense in an amber bottle.
zinc sulphate solution, 0.50 M $ZnSO_4$	- ~ 10 cm^3. Dissolve 14.5 g of $ZnSO_4 \cdot 7H_2O$ in water and make up to 100 cm^3.
silver residues bottle	- see below.

Silver residues. In view of the high cost of silver nitrate, we recommend
that all residues should be collected in a large amber bottle. Concentrated
hydrochloric acid should be added from time to time to ensure complete
precipitation as silver chloride. When the precipitate has settled, the
supernatant liquid may be poured off to make room for fresh residues. When
sufficient silver chloride has been collected in this way, it may be sold to
a chemical supplier, or recovery as silver nitrate performed in the laboratory.
(See a suitable laboratory manual for details.)

Experiment 46. Measuring the potential difference generated by some simple electrochemical cells

Requirements per student (or pair)	Notes
4 beakers, 50 cm^3	
emery paper	
6 filter paper strips	- small piece for cleaning electrodes
safety spectacles	
voltmeter, high resistance, 2 V	- most pH meters double as valve voltmeters. Alternatively, use a modern AVO meter with imput impedance 10^4 Ω.
3 connecting leads	- one end to be connected to voltmeter, the other to a metal strip via a crocodile clip.
copper, foil, Cu	- one strip, 5 cm x 0.5 cm.
silver, wire, Ag	- one piece, 5 cm.
zinc, foil, Zn	- one strip, 5 cm x 0.5 cm.
copper(II) sulphate solution, 1.0 M $CuSO_4$	- ~ 20 cm^3. Dissolve 250 g of $CuSO_4 \cdot 5H_2O$ in water and make up to 1000 cm^3.
potassium nitrate solution, saturated, 3 M KNO_3	- ~ 20 cm^3. Dissolve 300 g of solid in water and make up to 1000 cm^3.
silver nitrate solution, 0.10 M $AgNO_3$	- ~ 20 cm^3. See Experiment 45.
zinc sulphate solution, 1.0 M $ZnSO_4$	- ~ 20 cm^3. Dissolve 288 g of $ZnSO_4 \cdot 7H_2O$ in water and make up to 1000 cm^3.
silver residues bottle	- see Experiment 45

Experiment 47. Testing predictions about redox reactions

Requirements per student (or pair)	Notes
10 test-tubes, with corks or bungs to fit	
safety spectacles	
spatula	
test-tube rack	
copper, powder, Cu	– ~ 1 g.
1,1,1-trichloroethane, CH_3CCl_3	– ~ 2 cm^3.
zinc, powder, Zn	– ~ 1 g.
bromine water, Br_2(aq)	– ~ 6 cm^3. Carefully break a 1 cm^3 phial under 100 cm^3 of water and shake to dissolve.
chromium(III) chloride solution, 0.10 M $CrCl_3$	– ~ 3 cm^3. Dissolve 2.7 g of $CrCl_3 \cdot 6H_2O$ in water and make up to 100 cm^3.
iron(III) chloride solution, 0.10 M $FeCl_3$	– ~ 3 cm^3. Dissolve 1.6 g of $FeCl_3$ in water and make up to 100 cm^3.
potassium bromide solution, 0.10 M KBr	– ~ 3 cm^3. Dissolve 1.2 g of KBr in water and make up to 100 cm^3.
potassium chloride solution, 0.10 M KCl	– ~ 3 cm^3. Dissolve 0.8 g of KCl in water and make up to 100 cm^3.
potassium hexacyanoferrate(III) solution (ferricyanide), 0.10 M $K_3Fe(CN)_6$	– ~ 1 cm^3. Dissolve 3.3 g of $K_3Fe(CN)_6$ in water and make up to 100 cm^3.
potassium iodide solution, 0.10 M KI	– ~ 3 cm^3. Dissolve 1.7 g of KI in water and make up to 100 cm^3.
potassium manganate(VII) solution (permanganate), 0.020 M $KMnO_4$	– ~ 6 cm^3. Dissolve 0.32 g of $KMnO_4$ in water and make up to 100 cm^3.
potassium peroxodisulphate solution (persulphate), 0.10 M $K_2S_2O_8$	– ~ 3 cm^3. Dissolve 2.7 g of $K_2S_2O_8$ in water and make up to 100 cm^3.
sulphuric acid, dilute, 1 M H_2SO_4	– ~ 2 cm^3. Carefully add 55 cm^3 of concentrated acid, slowly and with constant stirring, to 800 cm^3 of water and make up to 1000 cm^3.

Experiment 48. Variation of cell potential with concentration

Requirements per student (or pair)	Notes
7 beakers, 50 cm^3	
6 filter paper strips	- ~ 10 cm x 1 cm.
safety spectacles	
voltmeter, high resistance, 2 V	- see Experiment 46.
2 connecting leads	- one end to voltmeter, the other with a crocodile clip.
copper, foil, Cu	- one strip, 5 cm x 0.5 cm.
silver, wire, Ag	- one piece, 5 cm.
copper(II) sulphate solution, 1.0 M $CuSO_4$	- ~ 20 cm^3. See Experiment 46.
potassium nitrate solution, 3 M KNO_3	- ~ 20 cm^3. See Experiment 46.
silver nitrate solutions (1) to (6): 0.10 M $AgNO_3$..........(1)	- ~ 20 cm^3. Dissolve 1.70 g of of $AgNO_3$ in distilled water and make up to 100 cm^3 in a graduated flask.
0.010 M $AgNO_3$.........(2)	- ~ 20 cm^3. Accurately dilute 1 volume of (1) with 9 volumes of water.
0.0033 M $AgNO_3$........(3)	- ~ 20 cm^3. Accurately dilute 1 volume of (2) with 2 volumes of water.
0.0010 M $AgNO_3$........(4)	- ~ 20 cm^3. Accurately dilute 1 volume of (2) with 9 volumes of water.
0.00033 M $AgNO_3$.......(5)	- ~ 20 cm^3. Accurately dilute 1 volume of (4) with 2 volumes of water.
0.00010 M $AgNO_3$.......(6)	- ~ 20 cm^3. Accurately dilute 1 volume of (4) with 9 volumes of water.
silver residues bottle	- see Experiment 45.

Experiment 49. Measuring the solubility products of some silver salts

Requirements per student (or pair)	Notes
5 beakers, 50 cm^3	
measuring cylinder, 25 cm^3	
emery paper or cloth	
6 filter paper strips	- ~ 10 cm x 1 cm.
safety spectacles	
voltmeter, high resistance	- see Experiment 46.
2 connecting leads	- one end to voltmeter, the other with a crocodile clip.
copper, foil, Cu	- one strip, 5 cm x 0.5 cm.
silver, wire, Ag	- one piece, 5 cm.
copper(II) sulphate solution, 1.0 M $CuSO_4$	- ~ 20 cm^3. See Experiment 46.
potassium bromide solution, 0.10 M KBr	- ~ 20 cm^3. See Experiment 47.
potassium chloride solution, 0.10 M KCl	- ~ 20 cm^3. See Experiment 47.
potassium iodate(V) solution, 0.10 M KIO_3	- ~ 20 cm^3. Dissolve 2.1 g of KIO_3 in water and make up to 100 cm^3.
potassium iodide solution, 0.10 M KI	- ~ 20 cm^3. See Experiment 47.
potassium nitrate solution, 3 M KNO_3	- ~ 20 cm^3. See Experiment 46.
silver nitrate solution, 0.10 M $AgNO_3$	- ~ 40 cm^3 or 10 cm^3. Consult the teacher.
silver residues bottle	- see Experiment 45.

Experiment 50. Reaction between sodium peroxide and water

<table>
<tr><td>

<u>Requirements per student (or pair)</u>

</td><td>

<u>Notes</u>

</td></tr>
</table>

Requirements per student (or pair)	Notes
3 test-tubes	
3 teat-pipettes	
Bunsen burner	
bench protection mat	
safety spectacles	
spatula	
test-tube rack	
wood splints (2 or 3)	
distilled water	
dilute sulphuric acid, 1 M H_2SO_4	- 2-3 cm^3. Carefully add 55 cm^3 of concentrated acid, slowly and with constant stirring, to 800 cm^3 of distilled water and make up to 1000 cm^3.
pentan-1-ol (amyl alcohol), $C_5H_{11}OH$	- 2-3 cm^3.
potassium dichromate solution, 0.02 M $K_2Cr_2O_7$	- 2-3 cm^3. Dissolve 0.5 g of solid in distilled water and make up to 100 cm^3.
sodium peroxide, Na_2O_2	- about 1 g. Fresh (the solid changes from yellow to white on exposure to air).
universal indicator solution	

Experiment 51. Heating the nitrates and carbonates of the S-block elements

<u>Requirements per student (or pair)</u>	<u>Notes</u>
30 test-tubes	
right-angled delivery tube, glass, with bung	- see diagram below.
2 Bunsen burners	
2 bench protection mats	
labels for test-tubes	
retort stand, clamp and boss	
safety spectacles	
2 spatulas	
2 stop-clocks or stop-watches	
2 test-tube holders	
2 test-tube racks	- students need <u>four</u> rows of 4 or 5 tubes each
wood splints	
nitrates and anhydrous carbonates of lithium, sodium, potassium, rubidium, caesium, magnesium, calcium, strontium and barium.	- a few grams of each available compound in the form of powder or fine crystals.
hydrochloric acid, dilute, 2 M HCl	- about 10 cm³. Add 172 cm³ of concentrated acid to 800 cm³ of distilled water and make up to 1000 cm³.
limewater	- about 50 cm³

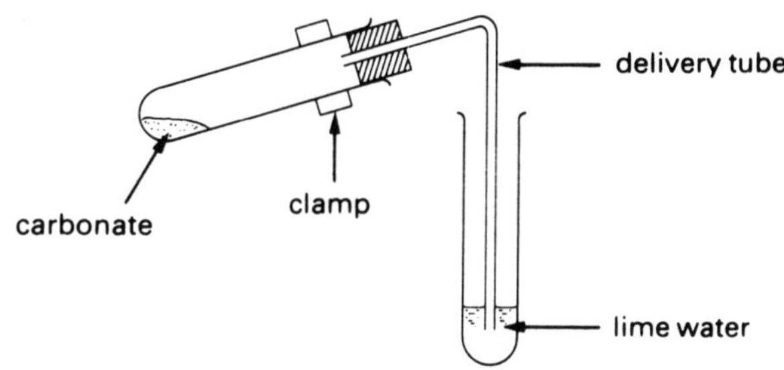

Experiment 52. The solubility of some salts of Group II elements

Requirements per student (or pair)	Notes

16 test-tubes

5 teat-pipettes marked at 1 cm^3 — draw in 1 cm^3 of water from a 10 cm^3 measuring cylinder and mark with a small blob of paint.

labels for test-tubes

safety spectacles

4 test-tube racks

distilled water

8 solutions labelled as follows: — a few cm^3 of each.

  0.1 M $Ba^{2+}(aq)$ — 26 g $Ba(NO_3)_2$ or 24 g $BaCl_2 \cdot 2H_2O$ per dm^3

  0.1 M $Ca^{2+}(aq)$ — 24 g $Ca(NO_3)_2 \cdot 4H_2O$ or 11 g $CaCl_2$ per dm^3

  0.1 M $Mg^{2+}(aq)$ — 26 g $Mg(NO_3)_2 \cdot 6H_2O$ or 20 g $MgCl_2 \cdot 6H_2O$ per dm^3

  0.1 M $Sr^{2+}(aq)$ — 21 g $Sr(NO_3)_2$ or 27 g $SrCl_2 \cdot 6H_2O$ per dm^3

  0.05 M $CO_3^{2-}(aq)$ — 5.3 g Na_2CO_3 (anhydrous) per dm^3

  1.0 M $OH^-(aq)$ — 40 g NaOH per dm^3 (freshly made)

  0.5 M $SO_3^{2-}(aq)$ — 53 g Na_2SO_3 or 126 g $Na_2SO_3 \cdot 7H_2O$ per dm^3

  0.5 M $SO_4^{2-}(aq)$ — 71 g Na_2SO_4 or 161 g $Na_2SO_4 \cdot 10H_2O$ per dm^3

Experiment 53. The solubility of the halogens in organic solvents

Requirements per student (or pair)	Notes
3 test-tubes, each with a bung or cork	
6 teat-pipettes, graduated at 1 cm^3	- draw in 1 cm^3 of water from a 10 cm^3 measuring cylinder and mark with a spot of paint.
safety spectacles	
test-tube rack	
3 reagent bottles for organic residues	- clearly labelled 'hexane residues', 'trichloroethane residues' and 'ethoxyethane residues'. See below for disposal.
bromine water	- about 5 cm^3. Carefully break a 1 cm^3 phial under 100 cm^3 of water and shake to dissolve.
chlorine water	- about 5 cm^3. Generate chlorine by adding concentrated hydrochloric acid dropwise to solid potassium manganate(VII) (permanganate) and bubble through water in a fume cupboard.
iodine solution, 0.01 M I_2 in KI(aq)	- about 5 cm^3. Dissolve 2 g of potassium iodide, KI, in 20 cm^3 of distilled water, add 1.25 g of iodine and stir to dissolve. Dilute to 100 cm^3.
ethoxyethane (diethyl ether), $(C_2H_5)_2O$	- about 10 cm^3.
hexane, C_6H_{14}, (or cyclohexane, C_6H_{12})	- about 10 cm^3.
1,1,1-trichloroethane, CH_3CCl_3	- about 10 cm^3.

Disposal of organic residues

Only the smallest quantities may be poured down a sink in a fume cupboard
with plenty of flowing water. It is better to collect residues and dispose
of them quickly by burning in a metal tray in the open away from people and
buildings. Alternatively, they can be allowed to evaporate from a metal
tray, covered with a mesh, on a roof with controlled access. Normally,
residues should not be stored, nor should residues from different experiments,
containing different chemicals, be mixed. For disposal of larger quantities,
consult your local authority.

Experiment 54. The action of dilute alkali on the halogens

Requirements per student (or pair)	Notes
5 test-tubes	
4 teat-pipettes	
safety spectacles	
test-tube rack	
wash-bottle of distilled water	
bromine water	- about 2 cm^3. Carefully break a 1 cm^3 phial under 100 cm^3 of distilled water and shake to dissolve.
iodine solution, 0.01 M I_2 in KI(aq)	- about 2 cm^3. Dissolve 2 g of potassium iodide, KI, in 20 cm^3 of distilled water, add 1.25 g of iodine and stir to dissolve. Dilute to 100 cm^3 with water.
sodium hydroxide solution, 2 M NaOH	- dissolve 80 g of sodium hydroxide pellets in distilled water and make up to 1000 cm^3.
sulphuric acid, dilute, 1 M H_2SO_4	- carefully add 55 cm^3 of concentrated sulphuric acid slowly and with constant stirring, to 800 cm^3 of distilled water and make up to 1000 cm^3.

Experiment 55. Halogen-halide reactions in aqueous solution

Requirements per student (or pair)	Notes
6 test-tubes, each with a bung or cork	
7 teat-pipettes	
safety spectacles	
reagent bottle for organic residues	- labelled 'hexane residues'.
test-tube rack	
bromine water	- about 5 cm^3. Carefully break a 1 cm^3 phial under 100 cm^3 of distilled water and shake to dissolve.
chlorine water	- about 5 cm^3. Generate chlorine by adding concentrated hydrochloric acid dropwise to solid potassium manganate(VII) (permanganate) and bubble through water in a fume cupboard.
hexane, C_6H_{14}, or cyclohexane, C_6H_{12}	- about 10 cm^3.
iodine solution, 0.01 M I_2 in KI(aq)	- about 5 cm^3. Dissolve 2 g of potassium iodide, KI, in 20 cm^3 of distilled water, add 1.25 g of iodine and stir to dissolve. Dilute to 100 cm^3.
potassium bromide solution, 0.1 M KBr	- about 5 cm^3. Dissolve 1.2 g of solid in distilled water and make up to 100 cm^3.
potassium chloride solution, 0.1 M KCl	- about 5 cm^3. Dissolve 0.75 g of solid in distilled water and make up to 100 cm^3.
potassium iodide solution, 0.1 M KI	- about 5 cm^3. Dissolve 1.7 g of solid in distilled water and make up to 100 cm^3.

Experiment 56. Reactions of solid halides

Requirements per student (or pair)	Notes
1 stirring rod, glass	
12 test-tubes	- preferably hard glass.
3 teat-pipettes	
Bunsen burner and protective mat	
filter paper strips	- for making test papers.
safety spectacles	
spatula	
test-tube holder	- for heating. Metal type preferred.
test-tube rack	
wash-bottle of distilled water	
ammonia solution, 0.880, NH_3	- about 2 cm^3, preferably in a dropping bottle.
hexane, C_6H_{14}, or cyclohexane, C_6H_{12}	- about 5 cm^3.
lead(II) ethanoate (acetate) solution, 0.1 M $(CH_3CO_2)_2Pb$	- about 5 cm^3. Dissolve 3.3 g of solid trihydrate in distilled water and make up to 100 cm^3.
manganese(IV) oxide, MnO_2	- about 2 g.
potassium bromide, KBr	- about 2 g.
potassium chloride, KCl	- about 2 g.
potassium iodide, KI	- about 2 g.
phosphoric(V) acid, 100%, H_3PO_4 (orthophosphoric acid)	- about 1 g. If the solid is not available, cautiously add a little phosphorus(V) oxide, P_4O_{10}, to 'syrupy' phosphoric acid (90%).
potassium dichromate(VI) solution, 0.1 M $K_2Cr_2O_7$	- about 5 cm^3. Dissolve 2.9 g in distilled water and make up to 100 cm^3.
starch solution, 1%	- about 5 cm^3. Mix 1 g of <u>soluble</u> starch to a paste with a <u>little</u> water. Add the paste to 100 cm^3 of boiling distilled water, with stirring, and boil for 1 minute. Cool. (Keep no longer than two weeks.)
sulphuric acid, concentrated, H_2SO_4	- about 2 cm^3.

Experiment 57. Reactions of halides in solution

<u>Requirements per student (or pair)</u>	<u>Notes</u>
18 test-tubes, 9 with corks or bungs	
8 teat-pipettes	
safety spectacles	
3 test-tube racks	
ammonia solution, 15 M NH_3	- about 15 cm^3. Dilute 33 cm^3 of 0.880 ammonia to 100 cm^3 with distilled water.
hexane, C_6H_{14}, or cyclohexane, C_6H_{12}	- about 5 cm^3.
hydrogen peroxide solution, H_2O_2	- about 5 cm^3. Dilute 30 cm^3 of '10-volume' hydrogen peroxide to 100 cm^3 with distilled water.
lead(II) nitrate solution, 0.1 M $Pb(NO_3)_2$	- about 5 cm^3. Dissolve 1.7 g in distilled water and make up to 100 cm^3.
nitric acid, dilute, 2 M HNO_3	- about 15 cm^3. Dilute 120 cm^3 of concentrated nitric acid to 1000 cm^3.
potassium bromide solution, 0.1 M KBr	- about 5 cm^3. Dissolve 1.2 g of solid in distilled water and make up to 100 cm^3.
potassium chloride solution, 0.1 M KCl	- about 5 cm^3. Dissolve 0.75 g of solid in distilled water and make up to 100 cm^3.
potassium iodide solution, 0.1 M KI	- about 5 cm^3. Dissolve 1.7 g of solid in distilled water and make up to 100 cm^3.
silver nitrate solution, 0.02 M $AgNO_3$	- about 10 cm^3. Dissolve 0.30 g of solid in distilled water and make up to 100 cm^3.
starch solution, 1%	- about 5 cm^3. Mix 1 g of <u>soluble</u> starch to a paste with a <u>little</u> water. Add the paste to 100 cm^3 of boiling distilled water, with stirring, and boil for 1 minute. Cool. (Keep for no longer than two weeks.)
sulphuric acid, dilute, 1 M H_2SO_4	- about 10 cm^3. Carefully add 55 cm^3 of concentrated sulphuric acid, slowly and with constant stirring, to 800 cm^3 of distilled water and make up to 1000 cm^3.

Experiment 58. Reactions of the halates

<u>Requirements per student (or pair)</u>	<u>Notes</u>
7 test-tubes	- preferably hard glass.
Bunsen burner and protective mat	
litmus papers, blue and red	
safety spectacles	
spatula	
test-tube holder	
test-tube rack	
wash-bottle of distilled water	
wood splints	
sodium chlorate(V), $NaClO_3$	- about 1 g.
sodium iodate(V), $NaIO_3$	- about 1 g.
cobalt(II) chloride solution, 0.01 M $CoCl_2$	- about 2 cm^3. Dissolve 0.25 g of solid $CoCl_2 \cdot 6H_2O$ in distilled water and make up to 100 cm^3 (or use $Co(NO_3)_2$).
potassium iodide solution, 0.1 M KI	- about 2 cm^3. Dissolve 1.7 g of solid in distilled water and make up to 100 cm^3.
sodium chlorate(I) solution, 1 M NaClO	- about 2 cm^3. Dilute the commercial solution (10-14%) with an equal volume of distilled water.
sulphuric acid, dilute, 1 M H_2SO_4	- about 5 cm^3. Carefully add 55 cm^3 of concentrated sulphuric acid, slowly and with constant stirring, to 800 cm^3 of distilled water and make up to 1000 cm^3.
universal indicator solution	- about 1 cm^3. Colour chart also needed.

Experiment 59. Balancing a redox reaction

Requirements per student (or pair)	Notes
2 burettes, 50 cm^3	
2 conical flasks, 250 cm^3	
2 funnels, small	- for filling burettes
2 measuring cylinders, 10 cm^3	
2 burette stands	
safety spectacles	
wash-bottle of distilled water	
white tile	
hydrochloric acid, 2 M HCl	- about 50 cm^3. Dilute 172 cm^3 of concentrated hydrochloric acid to 1000 cm^3 with distilled water.
potassium iodate(V) solution, 0.10 M KIO_3	- about 25 cm^3. Dissolve 21.40 g of A.R. KIO_3 in distilled water and make up to 1000 cm^3 in a volumetric flask.
potassium iodide solution, 1 M KI	- about 50 cm^3. Dissolve 166 g of solid in distilled water and make up to 1000 cm^3.
sodium thiosulphate solution, 0.10 M $Na_2S_2O_3$	- about 100 cm^3. Dissolve 24.8 g of A.R. $Na_2S_2O_3 \cdot 5H_2O$ in distilled water and make up to 1000 cm^3 in a volumetric flask.
starch solution, 0.2%	- about 10 cm^3. Mix 0.2 g of <u>soluble</u> starch to a paste with a little distilled water. Add the paste to 100 cm^3 of boiling water, with stirring, and boil for 1 minute. Cool. (Keep for no longer than a day or two.)

Experiment 60. Observation and deduction experiment

<u>Requirements per student (or pair)</u> | <u>Notes</u>

5 test-tubes
— preferably hard glass.

Bunsen burner and protective mat

filter paper strips
— for making test papers.

litmus papers, blue and red

safety spectacles

spatula

test-tube holder

test-tube rack

wash-bottle of distilled water

wood splints

potassium bromate(V), $KBrO_3$
— about 1 g. UNKNOWN - labelled 'F'.

chlorine water
— about 2 cm^3. Generate chlorine by adding concentrated hydrochloric acid dropwise to solid potassium manganate(VII) (permanganate) and bubble through water in a fume cupboard.

lead(II) ethanoate (acetate) solution, 0.1 M $(CH_3CO_2)_2Pb$
— about 2 cm^3. Dissolve 3.3 g of solid trihydrate in distilled water and make up to 100 cm^3.

nitric acid, dilute, 2 M HNO_3
— about 5 cm^3. Add 128 cm^3 of concentrated acid slowly to 800 cm^3 of distilled water and make up to 1000 cm^3.

silver nitrate solution, 0.02 M $AgNO_3$
— about 1 cm^3. Dissolve 0.30 g of solid in distilled water and make up to 100 cm^3.

Students may also ask for the following (issued only on request):

limewater, $Ca(OH)_2(aq)$
— about 2 cm^3 of saturated solution.

Experiment 61. Observation and deduction experiment

<u>Requirements per student (or pair)</u>	<u>Notes</u>

5 test-tubes — preferably hard glass.

Bunsen burner and protective mat

filter paper strips — for making test papers.

litmus papers, red and blue

safety spectacles

spatula

test-tube holder — preferably metal.

test-tube rack

wash-bottle of distilled water

wood splints

potassium iodide, KI — about 2 g. UNKNOWN - labelled 'D'.

potassium iodate(V), KIO_3 — about 2 g. UNKNOWN - labelled 'E'.

sulphuric acid, concentrated, H_2SO_4 — about 1 cm³.

potassium manganate(VII) solution, 0.01 M $KMnO_4$ (permanganate) — about 5 cm³. Dissolve 1.6 g of solid in water and make up to 1000 cm³.

sulphuric acid, dilute, 1 M H_2SO_4 — about 10 cm³. Carefully add 55 cm³ of concentrated sulphuric acid, slowly and with constant stirring, to 800 cm³ of distilled water and make up to 1000 cm³.

Students <u>may</u> also ask for the following (issued only on request):

ammonia solution, 0.880 NH_3 — about 2 cm³.

limewater, $Ca(OH)_2(aq)$ — about 2 cm³ of saturated solution.

hexane, C_6H_{14}, or cyclohexane, C_6H_{12} — about 2 cm³.

nitric acid, dilute, 2 M HNO_3 — about 5 cm³.

silver nitrate solution, 0.02 M $AgNO_3$ — about 1 cm³. Dissolve 0.30 g of solid in distilled water and make up to 100 cm³.

lead(II) ethanoate solution, 0.1 M $(CH_3CO_2)_2Pb$ — about 2 cm³. Dissolve 3.3 g of solid trihydrate in distilled water and make up to 100 cm³.

Experiment 62. Investigating the properties of Period 3 chlorides

Requirements per student (or pair)	Notes
2 measuring cylinders, 10 cm^3	- 1 must be dry.
4 teat-pipettes	
14 test-tubes	- 6 must be dry.
access to fume cupboard	
access to organic-residues bottle	- labelled 'hexane residues'.
gloves	- disposable type is adequate.
pH paper	- to cover range 1 to 7.
safety spectacles	
spatula	
test-tube rack	
thermometer, 0-100 °C ($\pm$1 °C)	
wash-bottle of distilled water	
aluminium chloride, $AlCl_3$	- about 2 g, preferably in the anhydrous form.
disulphur dichloride, S_2Cl_2	- about 2 cm^3.
hexane, C_6H_{14}	- about 35 cm^3.
magnesium chloride, $MgCl_2$	- about 2 g.
phosphorus trichloride, PCl_3	- about 2 cm^3.
silicon tetrachloride, $SiCl_4$	- about 2 cm^3 See below.
sodium chloride, $NaCl$	- about 2 g.
ammonia solution, 0.880 NH_3	- about 2 cm^3.
universal indicator solution	- about 5 cm^3. Colour chart also needed.

Silicon tetrachloride is readily hydrolysed and the product can firmly
seal a stopper in place and allow pressure to build up in the bottle.
Bottles containing silicon tetrachloride have been known to burst in store,
or on being opened. Bottles of $SiCl_4$ should be opened in the fume cupboard
with considerable caution after covering the bottle with a cloth. It is
particularly dangerous to return unused chemical to the bottle.

Disposal of organic residues

Only the smallest quantities may be poured down a sink in a fume cupboard
with plenty of flowing water. It is better to collect residues and dispose
of them quickly by burning in a metal tray in the open away from people and
buildings. Alternatively, they can be allowed to evaporate from a metal
tray, covered with a mesh, on a roof with controlled access. Normally,
residues should not be stored, nor should residues from different experiments,
containing different chemicals, be mixed. For disposal of larger quantities,
consult your local authority.

Experiment 63. Preparing anhydrous aluminium chloride

Requirements per student (or pair)	Notes
absorption tube	- straight form with one bulb. See diagram below for apparatus to be assembled by students.
access to desiccator	
adapter with 'T' connection	- with ground-glass-joints.
combustion tube	- 1 cm diameter or larger.
cylindrical dropping funnel, 50 cm³	- containing ground-glass-joint and tap.
glass rod	
pear-shaped flask, 50 cm³	- with ground-glass-joint.
receiver bottle with bung	- bung should be fitted with a short length of glass tubing to connect with the absorption tube. It should also have a hole to fit the combustion tube.
short length of glass tubing fitted into bung	- to make connection between combustion tube and adapter.
specimen (sample) tube	- fitted with lid.

access to balance capable of
  weighing to 0.01 g

access to fume cupboard

3 bosses, clamps and stands

Bunsen burner and bench mat

ceramic wool

forceps

long spatula

rubber tubing for connections

ruler, 30 cm

safety spectacles and disposable gloves

sticky labels

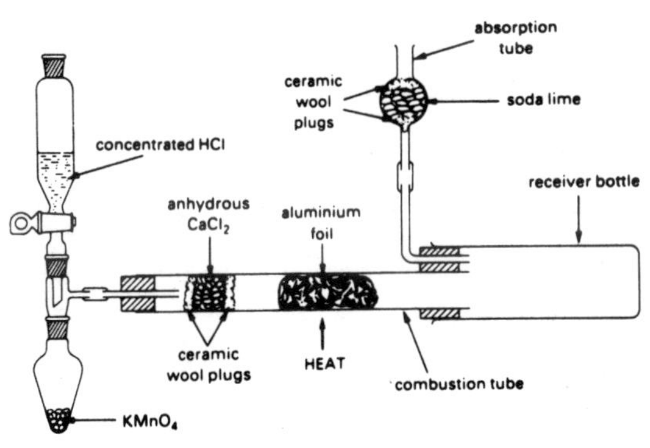

aluminium foil, Al	- about 0.25 g.
anhydrous calcium chloride, CaCl₂	- fused granular 3-8 mesh. About 20 g.
potassium manganate(VII) (permanganate), KMnO₄	- about 5 g.
hydrochloric acid, concentrated, HCl	- about 10 cm³.
soda-lime	- about 5 g.

Experiment 64. Investigating the properties of Period 3 oxides

Requirements per student (or pair)	Notes
Drechsel bottle	- fitted to SO_2 cylinder.
measuring cylinder, 10 cm^3	
measuring cylinder, 100 cm^3	- for bubbling SO_2 through water
teat-pipette	
6 test-tubes	
access to fume cupboard	
glass tubing	- with right-angled bend. One side should be long enough to reach the bottom of the 100 cm^3 measuring cylinder.
pH paper	
rubber tubing for connections	
safety spectacles	
spatula	
splints	
test-tube rack	
thermometer, 0-100 °C (±1 °C)	
wash-bottle of distilled water	
aluminium oxide, Al_2O_3	- about 2 g.
magnesium oxide, MgO	- about 2 g.
phosphorus(V) oxide, P_4O_{10}	- about 2 g.
silicon(IV) oxide, SiO_2	- about 2 g.
sodium peroxide, Na_2O_2	- about 2 g.
sulphur dioxide cylinder, fitted with a delivery tube	- or some other means of generating SO_2 gas.
universal indicator solution	- about 5 cm^3. Colour chart also needed.

Experiment 65. The reactions of tin and lead and their aqueous ions

Requirements per student (or pair)	Notes
15 test-tubes	
beaker, 250 cm^3	
3 teat-pipettes	
Bunsen burner and bench mat	
sticky labels for test-tubes	
safety spectacles	
disposable gloves	
test-tube holder	
2 test-tube racks	
wood splints	
universal indicator paper	- used to test gases.
lead, Pb	- 3 or 4 small pieces of foil.
tin, Sn	- 2 or 3 small pieces of granulated tin.
ammonia solution, 2 M NH$_3$	- about 20 cm^3. Add 112 cm^3 of 0.880 ammonia to water, and make up to 1000 cm^3.
hydrochloric acid, 2 M HCl	- about 20 cm^3. Add 172 cm^3 of concentrated acid to water and make up to 1000 cm^3.
hydrochloric acid, concentrated, HCl	- about 10 cm^3.
lead(II) nitrate solution, 0.1 M Pb(NO$_3$)$_2$	- about 20 cm^3. Dissolve 8.2 g of solid in distilled water and make up to 250 cm^3.
nitric acid, concentrated, HNO$_3$	- about 10 cm^3.
potassium chromate(VI) solution, 0.1 M K$_2$CrO$_4$	- about 10 cm^3. Dissolve 1.7 g of solid in distilled water and make up to 100 cm^3.
potassium iodide solution, 0.1 M KI	- about 5 cm^3. Dissolve 1.7 g of the solid in distilled water and make up to 100 cm^3.
potassium manganate(VII) solution in dilute ethanoic acid, 0.01 M KMnO$_4$ (permanganate)	- about 10 cm^3. Dissolve 0.16 g of solid in distilled water and make up to 100 cm^3 using about 10 cm^3 of 2 M ethanoic acid.
sodium hydroxide solution, 2 M NaOH	- about 20 cm^3. Carefully dissolve 80 g of sodium hydroxide pellets in water and make up to 1000 cm^3.
sodium sulphide solution, 0.02 M Na$_2$S	- about 1 cm^3. Dissolve 0.5 g of the solid in distilled water and make up to 100 cm^3. Use the fume cupboard - H$_2$S may be liberated.

Tin(II) chloride solution, 0.1 M SnCl$_2$. Carefully warm 5.6 g of solid SnCl$_2$·2H$_2$O with 50 cm^3 of concentrated hydrochloric acid until the solid dissolves. Add two small pieces of tin during the process to discourage oxidation to tin(IV). Make up to 250 cm^3 with distilled water, 20 cm^3 each.

Experiment 66. The preparations and reactions of tin(IV) oxide and lead(IV) oxide

Requirements per student (or pair)	Notes
conical flask, 250 cm^3	
glass rod	
2 measuring cylinders, 10 cm^3 and 100 cm^3	
suction filtration apparatus	- use a small Buchner funnel, with correct size filter paper, fitted to a Buchner flask. Attach to water pump.
2 specimen tubes	
3 teat-pipettes	
10 test-tubes	
Bunsen burner, tripod and gauze	
evaporating basin	
access to balance	
safety spectacles and gloves	
sticky labels	
spatula	
test-tube holder	
wash-bottle of distilled water	
wood splints	
blue litmus paper	
lead (small pieces), Pb	- about 6 g. Use foil if available.
lead(IV) oxide, PbO$_2$	- about 1 g.
tin (granulated), Sn	- about 2 g.
tin(IV) oxide, SnO$_2$	- about 1 g.
hydrochloric acid, dilute, 2 M HCl	- about 5 cm^3. See Experiment 65.
hydrochloric acid, concentrated, HCl	- about 5 cm.
nitric acid, dilute, 2 M HNO$_3$	- about 30 cm^3. Dilute 120 cm^3 of concentrated nitric acid to 1000 cm^3.
nitric acid, concentrated, HNO$_3$	- about 25 cm^3.
potassium iodide solution in dilute sulphuric acid, 1 M KI	- about 10 cm^3. Dissolve 16.5 g of the solid in water and make up to 100 cm^3 with distilled water and 10 cm^3 of 2 M H$_2$SO$_4$.
propanone, CH$_3$COCH$_3$ (acetone)	- about 10 cm^3.
sodium chlorate(I) solution, 1.5 M NaClO (hypochlorite)	- about 50 cm^3. Use the commercially available solution containing 10-14% chlorate(I).
sodium hydroxide solution, 2 M NaOH	- about 50 cm^3. See Experiment 65.
sodium hydroxide solution, 8 M NaOH	- about 5 cm^3. Carefully dissolve 32 g of sodium hydroxide pellets in 80 cm^3 of distilled water and make up to 100 cm^3.

Experiment 67. Observation and deduction exercise

Requirements per student (or pair)	Notes
boiling-tube	
filter funnel and filter paper	
10 test-tubes	
Bunsen burner and mat	
litmus paper, blue and red	
safety spectacles	
spatula	
test-tube holder	
test-tube rack	
wash-bottle of distilled water	
wood splints	
dilead(II) lead(IV) oxide, Pb_3O_4 (red lead)	- about 2 g. UNKNOWN - labelled Q.
lead(IV) oxide, PbO_2	- about 1 g. UNKNOWN - labelled I.
iron(III) chloride solution (neutralized), 0.3 M $FeCl_3$	- about 5 cm³. Dissolve 8.1 g in distilled water and make up to 100 cm³. The solution is neutralized with ammonia solution to the first permanent faint precipitate.
mercury(II) chloride solution, 0.1 M $HgCl_2$	- about 5 cm³. Dissolve 2.7 g in water and make up to 100 cm³.
nitric acid, 2 M HNO_3	- about 20 cm³. See Experiment 66.
potassium chromate(VI) solution, 0.5 M K_2CrO_4	- about 5 cm³. See Experiment 65.
potassium iodide solution, 0.1 M KI	- about 5 cm³. See Experiment 65.
potassium thiocyanate solution, 0.5 M KSCN	- about 2 cm³. Dissolve 5.0 g of solid in distilled water and make up to 100 cm³.
silver nitrate solution, 0.02 M $AgNO_3$	- about 2 cm³. Dissolve 0.30 g of solid in distilled water and make up to 100 cm³.
sodium hydroxide solution, 2 M NaOH	- about 5 cm³. See Experiment 65.
sulphuric acid, dilute 2 M H_2SO_4	- about 5 cm³. Carefully add 110 cm³ of concentrated sulphuric acid, slowly and with constant stirring to 800 cm³ of distilled water and make up to 1000 cm³.
tin(II) chloride solution, 0.25 M $SnCl_2$	- about 30 cm³. UNKNOWN - labelled H. See I4/1 but use 14 g of solid $SnCl_2 \cdot 2H_2O$ with 50 cm³ of concentrated hydrochloric acid and make up to 250 cm³ with distilled water.

Experiment 68. Illustrating the oxidation states of vanadium

<table>
<tr><td>

Requirements per student (or pair)

conical flask, 100 cm^3

filter funnel

measuring cylinder, 25 cm^3

5 test-tubes

Bunsen burner and bench mat

filter papers

safety spectacles and gloves

spatula

test-tube holder

test-tube rack

tripod and gauze

wash-bottle of distilled water

ammonium polytrioxovanadate
  (ammonium metavanadate), NH_4VO_3

sodium sulphite, Na_2SO_3

sulphuric acid, concentrated, H_2SO_4

zinc, dust, Zn

potassium iodide solution, 0.05 M KI

potassium manganate(VII) solution,
  (permanganate), 0.02 M $KMnO_4$

sodium thiosulphate solution,
  0.1 M $Na_2S_2O_3$

sulphuric acid, dilute, 1 M H_2SO_4

</td><td>

Notes

- ~ 1 g.

- ~ 1 g.

- ~ 10 cm^3.

- ~ 2 g. Small granules may be used, but dust is better.

- ~ 2 cm^3. Dissolve 8.3 g of solid in water and make up to 1000 cm^3.

- ~ 10 cm^3. Dissolve 3.2 g of solid in water and make up to 1000 cm^3.

- ~ 2 cm^3. Dissolve 25 g of the pentahydrate in water and make up to 1000 cm^3.

- ~ 25 cm^3. Carefully add 55 cm^3 of concentrated acid, slowly and with constant stirring, to 800 cm^3 of distilled water and make up to 1000 cm^3.

</td></tr>
</table>

Experiment 69. Illustrating the oxidation states of manganese

Requirements per student (or pair)	Notes
filter funnel	
stirring rod	
6 test-tubes	
3 filter papers	
safety spectacles and gloves	
spatula	
test-tube rack	
wash-bottle of distilled water	
manganese(IV) oxide, MnO_2 (manganese dioxide)	- ~ 1 g.
manganese(II) sulphate-4-water, $MnSO_4 \cdot 4H_2O$	- ~ 1 g. The mono-hydrate will serve.
sulphuric acid, concentrated, H_2SO_4	- ~ 1 cm³.
sodium hydroxide solution, 2 M NaOH	- ~ 10 cm³. Carefully dissolve 80 g of pellets in distilled water and make up to 1000 cm³.
potassium manganate(VII) solution, (permanganate), 0.01 M $KMnO_4$	- ~ 15 cm³. Dilute the solution made for Experiment 68.
sulphuric acid, dilute, 1 M H_2SO_4	- ~ 5 cm³. See Experiment 68.

Experiment 70. Relative stabilities of some complex ions

Requirements per student (or pair)	Notes
6 test-tubes	
test-tube rack	
wash-bottle of distilled water	
copper(II) sulphate solution, 0.20 M $CuSO_4$	- ~ 5 cm³. Dissolve 50 g of the pentahydrate in water and make up to 1000 cm³.
1,2-diaminoethane solution, 0.10 M $(CH_2NH_2)_2$	- ~ 25 cm³. Dissolve 6.0 g of solid in water and make up to 1000 cm³.
edta solution (disodium salt), 0.10 M $C_{10}H_{14}O_8N_2Na_2$	- ~ 25 cm³. Dissolve 37 g of the dihydrate in water and make up to 1000 cm³.
sodium chloride solution, 5 M NaCl	- ~ 25 cm³. Shake 300 g of solid with water and make up to 1000 cm³. This solution is almost saturated.
sodium ethanedioate (oxalate) solution, 0.20 M $(CO_2Na)_2$	- ~ 25 cm³. Dissolve 2.7 g of solid in water and make up to 1000 cm³.

Experiment 71. Determining the formula of a complex ion

<table>
<tr><td>

Requirements per student (or pair)

</td><td>

Notes

</td></tr>
<tr><td>

3 beakers, 100 cm^3

3 burettes, 50 cm^3

</td><td>

- these may be shared with other students. 10 cm^3 graduated pipettes, with safety fillers, may be substituted.

</td></tr>
<tr><td>

3 funnels, small

</td><td>

- for filling burettes (if used).

</td></tr>
<tr><td>

8 test-tubes, with bungs or corks

</td><td>

- optically matched, if possible, and fitting the colorimeter. If these tubes hold less than 15 cm^3, students will need 8 larger tubes <u>as well</u>.

</td></tr>
<tr><td>

3 burette stands

</td><td>

- if burettes are used.

</td></tr>
<tr><td>

colorimeter, with filters

safety spectacles

test-tube rack

wash-bottle of distilled water

ammonia solution, 0.10 M NH_3

</td><td>

- ~ 60 cm^3. See below.

</td></tr>
<tr><td>

ammonium sulphate solution, 2 M $(NH_4)_2SO_4$

</td><td>

- 50 cm^3. Dissolve 264 g of solid in water and make up to 1000 cm^3.

</td></tr>
<tr><td>

copper(II) sulphate solution, 0.10 M $CuSO_4$

</td><td>

- ~ 25 cm^3. See below.

</td></tr>
</table>

<u>Ammonia solution and copper(II) sulphate solution</u>. It is important for this experiment that these two solutions should be of <u>equal concentration</u>, but not necessarily 0.10 mol dm^{-3} precisely. We suggest that you do not rely on old stock solutions, but standardize the ammonia against a known acid. The copper sulphate solution can then be made to the <u>same</u> concentration.

If you need to make fresh ammonia solution, add 5.6 cm^3 of 0.880 ammonia to water, make up to 1000 cm^3 and mix thoroughly before standardization.

0.01 M $CuSO_4$ requires 25.0 g of the pentahydrate per 1000 cm^3. Adjust this concentration as necessary.

Experiment 72. Some redox chemistry of copper

Requirements per student (or pair)	Notes
beaker, 250 cm^3	
boiling-tube	
4 test-tubes	
Bunsen burner and bench mat	
safety spectacles	
spatula	
test-tube holder	- must fit boiling-tube
test-tube rack	
wash-bottle of distilled water	
copper, turnings, Cu	- ~ 1 g.
copper(II) chloride-6-water, $CuCl_2 \cdot 6H_2O$	- ~ 3 g. If this is not available, a mixture of sodium chloride and copper(II) sulphate may be used.
sodium chloride, NaCl	- ~ 3 g.
sodium sulphite, Na_2SO_3	- ~ 1 g. The hydrated form may be used.
sodium thiosulphate, $Na_2S_2O_3$	- ~ 1 g. The hydrated form may be used.
copper(II) sulphate solution, 0.1 M $CuSO_4$	- ~ 2 cm^3. See Experiment 71.
potassium iodide solution, 0.1 M KI	- ~ 4 cm^3. Dissolve 1.7 g of solid in water and make up to 100 cm^3.

Experiment 73. Investigating the use of cobalt(II) ions as a catalyst

Requirements per student (or pair)	Notes
beaker, 250 cm^3	
measuring cylinder, 25 cm^3	
stirring rod	
teat-pipette	
2 test-tubes	
Bunsen burner and bench mat	
safety spectacles	
spatula	
test-tube rack	
thermometer, 0-100 °C	
tripod and gauze	
wash-bottle of distilled water	
cobalt(II) chloride, $CoCl_2 \cdot 6H_2O$	- ~ 1 g.
hydrogen peroxide solution, 1.7 M H_2O_2	- ~ 20 cm^3 of '20-volume' H_2O_2, as purchased.
potassium sodium 2,3-dihydroxybutanedioate (tartrate), $CO_2K(CHOH)_2CO_2Na \cdot 4H_2O$	- ~ 1 g. Commonly known as 'Rochelle salt'.

Experiment 74. Catalysing the reaction between iodide ions and peroxodisulphate

Requirements per student (or pair)	Notes
4 beakers, 100 cm^3	
boiling-tube	
4 burettes, 50 cm^3	- may be shared with other students.
conical flask, 150 cm^3	
4 funnels, small	- for filling burettes.
4 test-tubes	
safety spectacles	
stop-clock	
test-tube rack	
wash-bottle of distilled water	
chromium(III) chloride solution, 0.1 M CrCl$_3$	- ~ 1 cm^3. Dissolve 3 g of the hexahydrate in water and make up to 100 cm^3.
iron(II) sulphate solution, 0.1 M FeSO$_4$	- ~ 1 cm^3. Dissolve 3 g of the heptahydrate in water and make up to 100 cm^3.
iron(III) chloride solution, 0.1 M FeCl$_3$	- ~ 1 cm^3. Dissolve 3 g of the hexahydrate in water with 2 cm^3 of conc. HCl and make up to 100 cm^3.
manganese(II) sulphate solution, 0.1 M MnSO$_4$	- ~ 1 cm^3. Dissolve 2.5 g of the tetrahydrate in water and make up to 100 cm^3.
potassium chromate(VI) solution, 0.1 M K$_2$CrO$_4$	- ~ 1 cm^3. Dissolve 2 g of solid in water and make up to 100 cm^3.
potassium manganate(VII) solution, (permanganate), 0.1 M KMnO$_4$	- ~ 1 cm^3. Dissolve 1.5 g of solid in water and make up to 100 cm^3.
potassium iodide solution, 0.2 M KI	- ~ 40 cm^3. Dissolve 33 g of solid in water and make up to 1000 cm^3.
potassium peroxodisulphate solution, (persulphate), 0.2 M K$_2$S$_2$O$_8$	- ~ 30 cm^3. Dissolve 54 g of solid in water and make up to 1000 cm^3.
sodium thiosulphate solution, 0.01 M Na$_2$S$_2$O$_3$	- ~ 50 cm^3. See Experiment 42.
sulphuric acid, dilute, 1 M H$_2$SO$_4$	- ~ 25 cm^3. See Experiment 68.
starch solution, 0.2%	- ~ 20 cm^3. Mix 0.2 g of 'soluble' starch into a paste with cold water and pour, with constant stirring, into 100 cm^3 of boiling distilled water.

The six 0.1 M solutions should be supplied in dropping bottles if possible. Otherwise, 6 teat-pipettes will be needed.

Experiment 75. Observation and deduction exercise

<u>Requirements per student (or pair)</u>	<u>Notes</u>
beaker, 100 cm^3	
3 boiling-tubes	
4 test-tubes and test-tube holder	
2 teat-pipettes	
watch-glass	
Bunsen burner and bench mat	
flame-test rod or wire	
litmus papers, red and blue	- or pH papers.
safety spectacles	
spatula	
test-tube rack	
wash-bottle of distilled water	
wood splints	
ammonium polytrioxovanadate, (metavanadate), NH_4VO_3	- ~ 0.5 g in a separate tube, labelled 'SUBSTANCE C'.
potassium dichromate(VI), $K_2Cr_2O_7$	- ~ 2 g in a separate tube, labelled 'SUBSTANCE D'.
copper, powder, Cu	- ~ 1 g.
hydrochloric acid, concentrated, HCl	- ~ 5 cm^3.
hydrogen peroxide, solution, 1.7 M H_2O_2	- ~ 5 cm^3 of '20 volume' H_2O_2
zinc, powder, Zn	- ~ 1 g.
potassium iodide solution, 0.6 M KI	- ~ 5 cm^3. Dissolve 10 g of solid in water and make up to 100 cm^3.
sodium hydroxide solution, 2 M NaOH	- ~ 50 cm^3. See Experiment 69.
sodium thiosulphate solution, 0.1 M $Na_2S_2O_3$	- ~ 10 cm^3. Dissolve 25 g of the pentahydrate in water and make up to 1000 cm^3.
sulphuric acid, dilute, 1 M H_2SO_4	- ~ 20 cm^3. See Experiment 68.

Experiment 76. Reactions of aluminium

Requirements per student (or pair)	Notes
beaker, 100 cm^3	
filter funnel	
10 test-tubes	
Bunsen burner and bench mat	
filter papers	
safety spectacles	
test-tube holder	
test-tube rack	
wash-bottle of distilled water	
aluminium foil, Al	– 6 pieces, 3 cm x 3 cm. Kitchen foil is suitable.
litmus papers, red and blue	
copper(II) chloride solution, 0.1 M CuCl$_2$	– ~ 10 cm^3. Dissolve 1.7 g of CuCl$_2 \cdot 2H_2O$ in water and make up to 100 cm^3.
hydrochloric acid, dilute, 2 M HCl	– ~ 10 cm^3. Dilute 172 cm^3 of concentrated hydrochloric acid to 1000 cm^3 with distilled water.
mercury(II) chloride solution, 0.1 M HgCl$_2$	– a few drops. Dissolve 2.7 g of solid in water and make up to 100 cm^3.
sodium carbonate solution, 1 M Na$_2$CO$_3$	– ~ 20 cm^3. Dissolve 106 g of anhydrous Na$_2$CO$_3$ in water and make up to 1000 cm^3.
sodium hydroxide solution, 2 M NaOH	– ~ 20 cm^3. Carefully dissolve 80 g of sodium hydroxide pellets in water and make up to 1000 cm^3.
sulphuric acid, dilute, 1 M H$_2$SO$_4$	– ~ 20 cm^3. Carefully add 55 cm^3 of concentrated sulphuric acid, slowly and with constant stirring, to 800 cm^3 of water and make up to 1000 cm^3.

Experiment 77. Anodizing aluminium

Requirements per student (or pair)	Notes

Requirements per student (or pair)   **Notes**

2 beakers, 100 cm³

Bunsen burner and bench mat

cotton-wool

forceps or tweezers

2 pins, large
       - heavy gauge, as used in optics experiments or entomology.

safety spectacles

thermometer, 0-100 °C

tripod and gauze

2 pieces of aluminium sheet
       - 7 cm x 3 cm. Mounted on wooden bars as shown below.

ammeter, 1 A

DC supply, 12 V

4 connecting leads
       - two must have crocodile clips at one end.

rheostat, 10 Ω, 4 A

sulphuric acid, 1 M H_2SO_4
       - ~ 75 cm³. See Experiment 76.

1,1,1-trichloroethane, CH_3CCl_3
       - ~ 2 cm³.

dye solution
       - a few drops or, if convenient, ~ 50 cm³. Dissolve 0.2 g of alizarin red or methyl violet in 100 cm³ or water and add 2 drops of 2 M NaOH. Congo red, and some other dyes, may be used but require 20% ethanol for solution. Standard indicator solutions are not sufficiently concentrated.

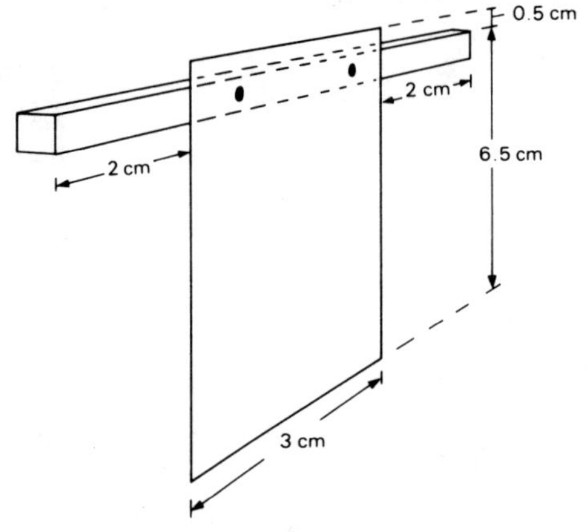

Experiment 78. Reactions of the oxoacids of nitrogen and their salts

Requirements per student (or pair)	Notes
beaker, 250 cm^3	
2 boiling-tubes	
6 test-tubes	
Bunsen burner and bench mat	
protective gloves	
safety spectacles	
spatula	
wash-bottle of distilled water	
wood splints	
aluminium foil, Al	– 3 cm square of kitchen foil.
copper turnings, Cu	– 4-5.
Devarda's alloy (Cu 50%, Al 45%, Zn 5%)	– ~ 1 g.
ice, crushed	– keep in freezer till requested.
iron(II) sulphate, $FeSO_4 \cdot 7H_2O$	– ~ 1 g.
magnesium ribbon, Mg	– ~ 5 cm.
sodium nitrite, $NaNO_2$	– ~ 3 g.
nitric acid, concentrated, 16 M HNO_3	– ~ 4 cm^3.
nitric acid, dilute, 2 M HNO_3	– ~ 10 cm^3. Carefully add 128 cm^3 of concentrated nitric acid to 800 cm^3 of water and make up to 1000 cm^3.
potassium iodide solution, 0.5 M KI	– ~ 1 cm^3. Dissolve 8.3 g of KI crystals in water and make up to 100 cm^3.
potassium manganate(VII) solution, 0.2 M $KMnO_4$	– ~ 1 cm^3. Dissolve 3.2 g of $KMnO_4$ crystals in water and make up to 100 cm^3.
sodium hydroxide solution, 2 M NaOH	– ~ 15 cm^3. See Experiment 76.
sulphuric acid, dilute, 1 M H_2SO_4	– ~ 20 cm^3. See Experiment 76.

Experiment 79. Investigating some reactions of the oxo-salts of sulphur

Requirements per student (or pair)	Notes
bottle labelled 'silver residues'	
filter funnel	
10 test-tubes & 1 boiling-tube	
Bunsen burner and bench mat	
filter papers and test strips	– strips are for testing gases.
safety spectacles and protective gloves	
spatula	
test-tube holder	
test-tube rack	
wash-bottle of distilled water	
wood splints	
sodium peroxodisulphate, $Na_2S_2O_8$ (persulphate)	– ~ 1 g. The potassium salt will serve.
sodium sulphate-10-water, $Na_2SO_4 \cdot 10H_2O$	– ~ 1 g.
sodium sulphite-7-water, $Na_2SO_3 \cdot 7H_2O$	– ~ 1 g.
sodium thiosulphate-5-water, $Na_2S_2O_3 \cdot 5H_2O$	– ~ 1 g.
sulphur, finely powdered roll, S	– ~ 1 g. Preferred to 'flowers'.
hydrochloric acid, dilute, 2 M HCl	– ~ 15 cm^3. See Experiment 76.
iodine solution, 0.2 M I_2 in KI(aq)	– 10 cm^3. Dissolve 5.1 g of iodine and 10 g of KI crystals in water and make up to 100 cm^3.
iron(III) chloride solution, 0.5 M $FeCl_3$	– 5 cm^3. Dissolve 13 g of solid in water with 2 cm^3 of concentrated hydrochloric acid and make up to 100 cm^3.
potassium dichromate(VI) solution, 0.1 M $K_2Cr_2O_7$	– ~ 2 cm^3. Dissolve 2.9 g of solid in water and make up to 100 cm^3.
potassium iodide solution, 0.5 M KI	– ~ 5 cm^3. See Experiment 78.
silver nitrate solution, 0.1 M $AgNO_3$	– ~ 10 cm^3. Dissolve 1.7 g of solid in water and make up to 100 cm^3. Keep in amber bottles.
sodium hydroxide solution, 2 M NaOH	– ~ 10 cm^3. See Experiment 76.
sodium peroxodisulphate solution, 0.2 M $Na_2S_2O_8$ (persulphate)	– ~ 5 cm^3. 4.8 g $Na_2S_2O_8$ / 5.4 g $K_2S_2O_8$
sodium sulphate solution, 0.2 M Na_2SO_4	– ~ 5 cm^3. 6.5 g $Na_2SO_4 \cdot 10H_2O$
sodium sulphite solution, 0.2 M Na_2SO_3	– ~ 15 cm^3. 5.0 g $Na_2SO_3 \cdot 7H_2O$
sodium thiosulphate solution, 0.2 M $Na_2S_2O_3$	– ~ 5 cm^3. 5.0 g $Na_2S_2O_3 \cdot 5H_2O$

The four sodium solutions are made up per 100 cm^3.

Experiment 80. Observation and deduction exercise

Requirements per student (or pair)	Notes
beaker, 100 cm^3	
2 teat-pipettes	
evaporating dish	
flame-test wire or rod	- platinum wire or silica rod.
4 test-tubes and boiling-tube	
watch-glass	
Bunsen burner and bench mat	
safety spectacles	
strips of filter paper	- for testing gases.
test-tube rack	
tripod and gauze	
wash-bottle of distilled water	
litmus papers, red and blue	
iron(II) sulphate	- ~ 2 g.
ammonium ethanoate (acetate) solution, 3 M CH$_3$CO$_2$NH$_4$	- ~ 10 cm^3. Dissolve 23 g of solid in water and make up to 100 cm^3.
barium chloride solution, 0.5 M BaCl$_2$	- ~ 2 cm^3. Dissolve 12 g of crystals in water and make up to 100 cm^3.
copper(II) sulphate solution, 0.5 M CuSO$_4$	- ~ 5 cm^3. Dissolve 12 g of hydrate in water and make up to 100 cm^3.
hydrochloric acid, concentrated, 12 M HCl	- ~ 5 cm^3.
hydrochloric acid, dilute, 2 M HCl	- ~ 15 cm^3. See Experiment 76.
lead(II) ethanoate (acetate) solution, 0.25 M (CH$_3$CO$_2$)$_2$Pb	- ~ 2 cm^3. Dissolve 9.5 g of hydrate in water and make up to 100 cm^3.
potassium dichromate(VI) solution, 0.5 M K$_2$Cr$_2$O$_7$	- ~ 2 cm^3. Dissolve 14 g of solid in water and make up to 100 cm^3.
potassium iodide solution, 0.5 M KI	- ~ 10 cm^3. See Experiment 78.
potassium thiocyanate solution, 0.5 M KSCN	- ~ 2 cm^3. Dissolve 5 g of solid in water and make up to 100 cm^3.
silver nitrate solution, 0.1 M AgNO$_3$	- ~ 4 cm^3. See Experiment 79.
sodium chloride solution, 0.1 M NaCl	- ~ 4 cm^3. Dissolve 0.6 g of solid in water and make up to 100 cm^3.
sodium peroxodisulphate (persulphate) solution, 0.05 M Na$_2$S$_2$O$_8$, labelled 'SOLUTION D'	- ~ 10 cm^3. Dissolve 1.2 g of Na$_2$S$_2$O$_8$ (or 1.4 g of K$_2$S$_2$O$_8$) in water and make up to 100 cm^3.
sodium sulphate solution, 0.1 M Na$_2$SO$_4$ labelled 'SOLUTION B'	- ~ 15 cm^3. Dissolve 3.4 g of Na$_2$SO$_4 \cdot$10H$_2$O in water and make up to 100 cm^3.
sodium thiosulphate solution, 0.1 M Na$_2$S$_2$O$_3$, labelled 'SOLUTION C'	- ~ 40 cm^3. Dissolve 22 g of Na$_2$S$_2$O$_3 \cdot$5H$_2$O in water and make up to 1000 cm^3.
sulphuric acid, dilute, 1 M H$_2$SO$_4$	- ~ 20 cm^3. See Experiment 76.

Experiment 81. Chemical properties of alkanes

Requirements per student (or pair)	Notes
5 test-tubes, dry, with corks to fit	
1 test-tube, dry, covered with foil	- aluminium foil should be tightly wrapped round the tube to prevent light from entering.
watch-glass, hard glass	
access to fume cupboard	
test-tube rack	
Bunsen burner and bench mat	
gloves	- disposable type is adequate.
safety spectacles	
lamp with 100 watt bulb	- room fitting or table lamp.
wood splints	
ammonia solution, 0.880	- small bottle with glass stopper.
bromine dissolved in 1,1,1-trichloro-ethane (with teat-pipette)	- about 2 cm³. See note below.
bromine water (with teat-pipette)	- about 2 cm³. See note below.
cyclohexane	- about 6 cm³.
potassium manganate(VII) solution (permanganate), 0.01 M KMnO₄	- about 2 cm³. Dissolve about 1 g in distilled water and make up to 1000 cm³
sulphuric acid, concentrated	- about 2 cm³. Small bottle with dropper.
sulphuric acid, dilute, 1 M H₂SO₄	- about 2 cm³.

Bromine water

This should be prepared in a fume cupboard and protective gloves should be worn. Use 1 cm³ of bromine in 125 cm³ of water. It is probably safest to crush an ampoule using a pair of pliers but care must be taken when cleaning out the vessel after use. This solution should be kept in a well stoppered amber bottle in a ventilated store.

Bromine dissolved in 1,1,1-trichloroethane

Prepared as for bromine water. Instead of water, use 1,1,1-trichloroethane. This solution should not be kept longer than a week. After a few days, some gas may build up over the solution so, before handing it over to the students, release the stopper in a fume cupboard.

Experiment 82. Chemical properties of alkenes

The requirements are exactly the same as for Experiment 81 except that cyclohexene is used instead of cyclohexane.

Experiment 83. Hydrolysing organic halogen compounds

Requirements per student (or pair)	Notes

Requirements per student (or pair)

beaker, 250 cm^3

measuring cylinder, 10 cm^3

thermometer, 0-100 °C

5 test-tubes fitted with corks

Bunsen burner, tripod, gauze
  and bench protection mat

safety spectacles

clock with second hand

protective gloves

test-tube rack

waterproof marker or chinagraph pencil

1-bromobutane, C_4H_9Br

1-chlorobutane, C_4H_9Cl

1-iodobutane, C_4H_9I

chlorobenzene, C_6H_5Cl

ethanol, C_2H_5OH

silver nitrate solution, 0.05 M $AgNO_3$

Notes

- the teacher may ask instead for
  a thermostatically controlled
  water-bath.

- room clock or wrist-watches are
  adequate.
- disposable type preferred.

- ~ 1 cm^3 of each. Teat-pipettes also
  required - see introductory notes
  for technicians.

- ~ 20 cm^3. Industrial methylated
  spirit (IMS 74 o.p.) is
  suitable.

- ~ 5 cm^3. Dissolve 2.1 g of solid
  in distilled water and make up to
  250 cm^3. Store and dispense in
  amber bottles.

Experiment 84. Preparing a halogeno-alkane

Requirements per student (or pair)	Notes

conical flask, 100 cm³ with bung

ground-glass-joint apparatus shown in diagram below
— 14/23 joints and 50 cm³ flask preferred.

measuring cylinder, 25 cm³
— to fit apparatus in diagram below.

rubber tubing
— to fit condenser and receiver in diagram below.

separating funnel, 50 cm³, with stopper

thermometer, 0-100 °C

anti-bumping chips

Bunsen burner, tripod and gauze

3 retort stands, bosses and clamps

safety spectacles and protective gloves

spatula

2-methylpropan-2-ol, $(CH_3)_3COH$, (tert-butyl alcohol)
— ~ 9 cm³. May need warming (M.Pt. 25 °C) - since students need it in the liquid state.

sodium sulphate, anhydrous, Na_2SO_4
— ~ 5 g.

hydrochloric acid, concentrated, HCl
— ~ 20 cm³.

sodium hydrogencarbonate solution, saturated, $NaHCO_3$
— ~ 10 cm³. Shake about 9 g of solid in 100 cm³ of water. Decant from excess solid after standing.

balance, sensitivity ± 0.1 g

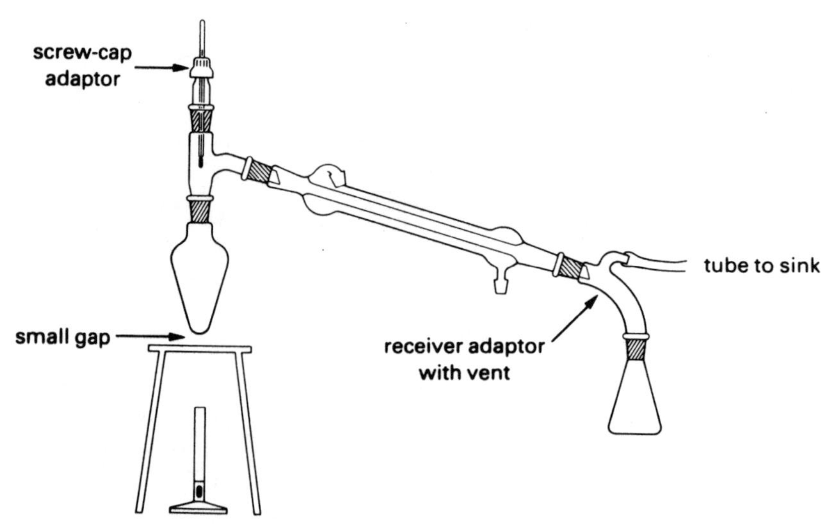

Experiment 85. Chemical properties of ethanol (Parts A & B)

Requirements per student (or pair)	Notes

A. test-tube

safety spectacles

wash-bottle of distilled water

ethanol — ~ 1 cm³. I.M.S. is suitable.

universal indicator solution — in dropping bottle.

B. beaker, 250 cm³

boiling-tube

funnel, small, wide stem — for adding solid to flask.

ground-glass-joint apparatus — as shown below. Rubber tubing also required for condenser and receiver.

measuring cylinder, 10 cm³

3 teat-pipettes

test-tube and holder

balance, sensitivity ± 0.1 g

Bunsen burner, tripod, gauze and bench mat

2 retort stands, bosses and clamps

safety spectacles and gloves

spatula

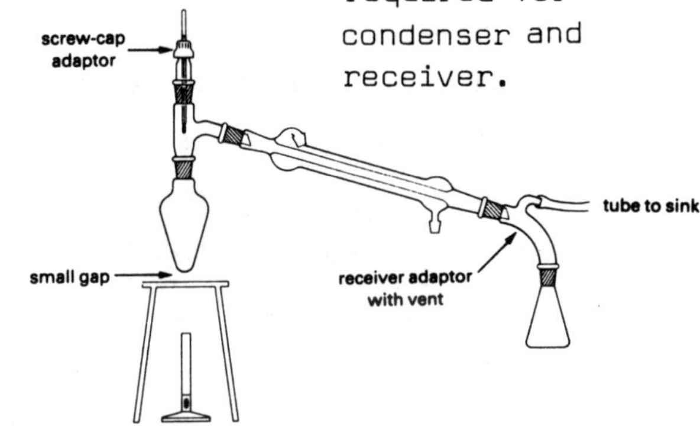

ethanol, C_2H_5OH — ~ 5 cm³. I.M.S. is suitable.

sodium dichromate(VI), $Na_2Cr_2O_7$ — ~ 3 g.

ammonia solution, 2 M NH_3 — ~ 2 cm³. Carefully dilute 45 cm³ of 0.880 ammonia to 1000 cm³ with water.

anti-bumping granules

Fehling's solutions, 1 and 2 — ~ 1 cm³ of each. See below.

silver nitrate solution, 0.05 M $AgNO_3$ — ~ 5 cm³. See Experiment 83.

sodium hydroxide solution, 2 M NaOH — ~ 1 cm³. Carefully dissolve 80 g of sodium hydroxide pellets in water and make up to 1000 cm³.

sulphuric acid, dilute, 1 M H_2SO_4 — ~ 10 cm³. Carefully add 55 cm³ of concentrated sulphuric acid, slowly and with constant stirring, to 800 cm³ of water and make up to 1000 cm³.

<u>Fehling's solution 1.</u> Dissolve 7 g of copper(II) sulphate crystals in water containing a few drops of dilute sulphuric acid, and make up to 100 cm³.

<u>Fehling's solution 2.</u> Dissolve 12 g of sodium hydroxide pellets and 34 g of sodium potassium 2,3-dihydroxybutanedioate (tartrate), 'Rochelle salt', in water. Filter if necessary and make up to 100 cm³.

Experiment 85.   Chemical properties of ethanol (Parts C, D & E)

<u>Requirements per student (or pair)</u> | Notes

C. funnel, small, wide stem

ground-glass-joint apparatus

measuring cylinder, 10 cm³

teat-pipette

test-tube

balance, sensitivity ± 0.1 g

Bunsen burner, tripod, gauze
   and bench mat

2 retort stands, bosses and clamps

safety spectacles and gloves

spatula

universal indicator papers

anti-bumping granules

ethanol, C_2H_5OH

sodium carbonate, Na_2CO_3

sodium dichromate(VI), $Na_2Cr_2O_7$

sulphuric acid, concentrated, H_2SO_4

sulphuric acid, dilute, 1 M H_2SO_4

D. 2 teat-pipettes

test-tube

safety spectacles

ethanol, C_2H_5OH

iodine solution, 10% I_2 in KI(aq)

sodium hydroxide solution, 2 M NaOH

E. test-tube

watch-glass

filter papers

forceps

safety spectacles

test-tube rack

wood splint

ethanol, C_2H_5OH

sodium metal, Na

Notes

- as shown below. Rubber tubing
  also required for condenser and
  receiver.

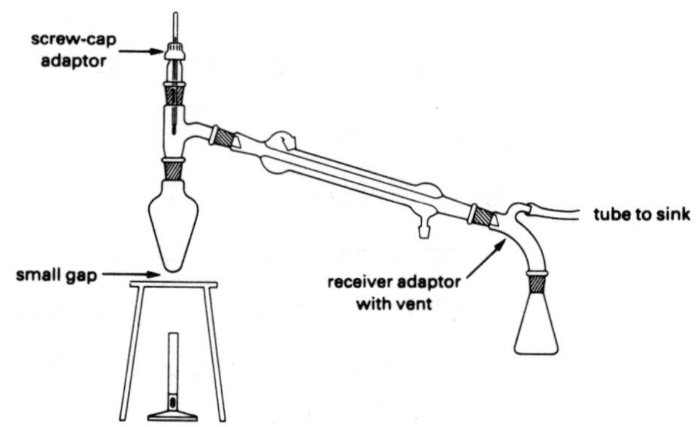

- ~ 1 cm³. I.M.S. is suitable.

- ~ 1 g.   (anhydrous)

- ~ 5 g.

- ~ 2 cm³.

- ~ 10 cm³. See parts A and B.

- ~ 1 cm³. I.M.S. is suitable.

- ~ 1 cm³. Dissolve 10 g of iodine
  and 20 g of potassium iodide in
  water and make up to 100 cm³.

- ~ 1 cm³. See parts A and B.

- for removing oil from sodium.

- for handling sodium.

- ~ 1 cm³. I.M.S. is suitable.

- two 1 mm cubes under the same
  oil as is used for storage.

Experiment 85. Chemical properties of ethanol (Parts F & G)

Requirements per student (or pair)	Notes
F. beaker, 100 cm³	
teat-pipette	
test-tube	
Bunsen burner and bench mat	
safety spectacles	
test-tube holder	
ethanol, C_2H_5OH	– ~ 2 cm³
ethanoic (acetic) acid, glacial, CH_3CO_2H	– ~ 1 cm³
sulphuric acid, concentrated, H_2SO_4	– ~ 1 cm³
sodium carbonate solution, 1 M Na_2CO_3	– ~ 20 cm³. Dissolve 106 g of anhydrous Na_2CO_3 in water and make up to 1000 cm³.
G. conical flask, 100 cm³, fitted with delivery tubes	– see diagram below.
teat-pipette	
3 test-tubes with corks	
water-trough or large beaker	– see diagram below.
Bunsen burner and bench mat	
ceramic wool	– see diagram below.
2 retort stands, bosses and clamps	
safety spectacles	
ethanol, C_2H_5OH	– ~ 2 cm³
pumice stone, 4–8 mesh	– see diagram below.
bromine water, Br_2(aq)	– ~ 1 cm³. See parts C, D and E.
potassium manganate(VII) (permanganate) solution, 0.01 M $KMnO_4$	– ~ 1 cm³. Dissolve 1.6 g of solid in water and make up to 100 cm³.

<u>Bromine water.</u> This should be prepared in a fume cupboard. Use 1 cm³ of bromine in 125 cm³ of water. The safest procedure is to crush an ampoule under water using pliers. Store the solution in an amber bottle in a fume cupboard.

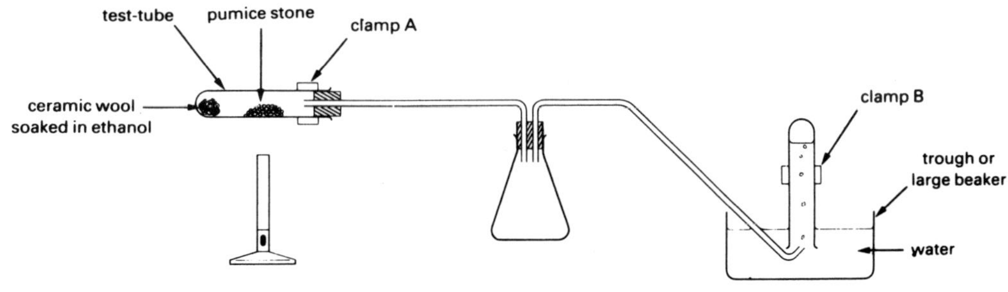

Experiment 86. Chemical properties of phenol

Requirements per student (or pair)	Notes
2 beakers, 250 cm^3	
teat-pipette	
6 test-tubes, with corks	
thermometer, 0-100 $^\circ$C	
Bunsen burner, tripod, gauze and bench mat	
filter papers	- for removing oil from sodium.
forceps	- for handling sodium.
safety spectacles and gloves	- disposable gloves preferred.
spatula	
test-tube holder	
test-tube rack	
wash-bottle of distilled water	
wood splints	
hydrochloric acid, concentrated, HCl	- ~ 2 cm^3
phenol, C$_6$H$_5$OH	- ~ 12 g
sodium metal, Na	- two 1 mm cubes, under the same oil as is used for storage.
sodium hydrogencarbonate, NaHCO$_3$	- ~ 1 g.
universal indicator solution	- in dropping bottle.
bromine water, Br$_2$(aq)	- ~ 1 cm^3. See Experiment 85 (F and G).
iron(III) chloride solution, 0.5 M FeCl$_3$	- ~ 1 cm^3. Dissolve 13 g of solid in water with 2 cm^3 of concentrated hydrochloric acid and make up to 100 cm^3.
sodium carbonate solution, 1 M Na$_2$CO$_3$	- ~ 2 cm^3. See Experiment 85 (F and G).
sodium hydroxide solution, 2 M NaOH	- ~ 5 cm^3. See Experiment 85 (F and G).

Experiment 87. Reactions of amines

Requirements per student (or pair)	Notes

6 beakers, 100 cm^3

2 beakers, 600 cm^3

4 boiling-tubes, with corks

stirring rod

3 teat-pipettes

6 test-tubes, with corks

thermometer, 0-100 °C

3 watch-glasses

Bunsen burner, tripod, gauze
  and bench mat

labels for test-tubes and beakers

protective gloves — disposable type preferred.

safety spectacles

spatula

test-tube rack

wood splint

crushed ice — keep in freezer till asked for.

wash-bottle of distilled water

universal indicator paper

ammonium chloride, NH_4Cl — ~ 6 g.

butylamine, $C_4H_9NH_2$ — ~ 5 cm^3.

hydrochloric acid, concentrated, HCl — ~ 5 cm^3.

naphthalen-2-ol (2-naphthol), $C_{10}H_7OH$ — ~ 1 g.

phenol, C_6H_5OH — ~ 1 g.

phenylamine (aniline) $C_6H_5NH_2$ — ~ 5 cm^3.

sodium chloride, NaCl — ~ 20 g. Commercial salt will do.

sodium nitrite, $NaNO_2$ — ~ 2 g.

ammonia solution, 2 M NH_3 — ~ 5 cm^3. See Experiment 85 (A and B).

bromine water, Br_2(aq) — ~ 2 cm^3. See Experiment 85 (A and B).

copper(II) sulphate solution, 1 M $CUSO_4$ — ~ 6 cm^3. Dissolve 25 g of copper(II) sulphate crystals in water and make up to 100 cm^3.

limewater, saturated, $Ca(OH)_2$ — ~ 2 cm^3. See below.

sodium hydroxide solution, 2 M NaOH — ~ 20 cm^3. See Experiment 85 (A and B).

Limewater. Shake 75 g of calcium hydroxide in 2 dm^3 of water in a Winchester bottle and allow to settle. Siphon off as required and refill with water.

Experiment 88. Observation and deduction exercise

Requirements per student (or pair)	Notes
2 beakers, 100 cm³	
2 beakers, 250 cm³	
6 test-tubes	
Bunsen burner, tripod, gauze and bench mat	
protective gloves	
safety spectacles	
spatula	
test-tube holder	
test-tube rack	
wash-bottle of distilled water	
crushed ice	- keep in freezer till asked for.
2 wood splints	
ethanol, C_2H_5OH	- ~ 15 cm³. UNKNOWN - labelled 'D'.
ethanoic (acetic) acid, glacial, CH_3CO_2H	- ~ 15 cm³. UNKNOWN - labelled 'C'.
hydrochloric acid, concentrated, HCl	- ~ 1 cm³.
phenylammonium chloride (aniline hydrochloride), $C_6H_5NH_3Cl$	- ~ 2 g. UNKNOWN - labelled 'E'.
phenol, C_6H_5OH	- ~ 1 g.
phosphorus pentachloride, PCl_5	- ~ 2 g.
potassium iodide, KI	- ~ 1 g.
sodium carbonate, anhydrous, Na_2CO_3	- ~ 5 g.
sodium nitrite, $NaNO_2$	- ~ 2 g.
sulphuric acid, concentrated, H_2SO_4	- ~ 2 cm³.
ammonia solution, 2 M NH_3	- ~ 5 cm³. See Experiment 85 (A and B).
bromine water, Br_2(aq)	- ~ 2 cm³. See Experiment 85 (F and G).
limewater, saturated, $Ca(OH)_2$	- ~ 2 cm³. Issue only on request.
nitric acid, dilute, 2 M HNO_3	- ~ 5 cm³. Carefully add 128 cm³ of concentrated nitric acid to water and make up to 1000 cm³.
silver nitrate solution, 0.05 M $AgNO_3$	- ~ 5 cm³. See Experiment 83.
sodium carbonate solution, 1 M Na_2CO_3	- ~ 20 cm³. See Experiment 85 (F and G).
sodium chlorate(I) solution (hypochlorite), NaClO	- ~ 5 cm³. 10-14% available chlorine as purchased. Commercial bleach is probably adequate.
sodium hydroxide solution, 2 M NaOH	- ~ 2 cm³. See Experiment 85 (A and B).
sulphuric acid, dilute, 1 M H_2SO_4	- ~ 20 cm³. See Experiment 85 (A and B).
potassium dichromate(VI) solution, 0.1 M $K_2Cr_2O_7$	- ~ 5 cm³. Dissolve 2.9 g of solid in water and make up to 100 cm³.

Experiment 89. Reactions of aldehydes and ketones

Requirements per student (or pair)	Notes
beaker, 250 cm^3	
6 test-tubes	
Bunsen burner, tripod, gauze and bench protection mat	
protective gloves	- disposable type preferred.
safety spectacles	
ethanal (acetaldehyde), CH_3CHO	- ~ 10 cm^3.
propanone (acetone), CH_3COCH_3	- ~ 10 cm^3.
ammonia solution, 2 M NH_3	- ~ 5 cm^3. Carefully dilute 45 cm^3 of 0.880 ammonia to 1000 cm^3 with water
2,4-dinitrophenylhydrazine solution, $C_6H_3(NO_2)_2NHNH_2$	- ~ 5 cm^3. Dissolve 0.5 g of solid in 25 cm^3 of methanol and add, dropwise, 1 cm^3 of concentrated sulphuric acid. Filter off any undissolved solid.
Fehling's solutions, 1 and 2	- ~ 3 cm^3 of each. See below.
iodine solution, 10% I_2 in KI(aq)	- ~ 1 cm^3. Dissolve 10 g of iodine and 20 g of potassium iodide in water and make up to 100 cm^3.
potassium dichromate(VI) solution, 0.1 M $K_2Cr_2O_7$	- ~ 2 cm^3. Dissolve 2.9 g of solid in water and make up to 100 cm^3.
silver nitrate solution, 0.05 M $AgNO_3$	- ~ 3 cm^3. Dissolve 2.1 g of solid in distilled water and make up to 250 cm^3. Store and dispense in amber bottles.
sodium hydrogensulphite solution, saturated, $NaHSO_3$	- ~ 5 cm^3. Shake 40 g of sodium metabisulphite, $Na_2S_2O_5$, in 100 cm^3 of water. Decant from excess solid after standing.
sodium hydroxide solution, 2 M NaOH	- ~ 10 cm^3. Dissolve 80 g of pellets in water and make up to 1000 cm^3.
sulphuric acid, dilute, 1 M H_2SO_4	- ~ 5 cm^3. Carefully add 55 cm^3 of concentrated sulphuric acid, slowly and with constant stirring, to 800 cm^3 of water and make up to 1000 cm^3.

Fehling's solution 1. Dissolve 7 g of copper(II) sulphate crystals in water containing a few drops of dilute sulphuric acid, and make up to 100 cm^3.

Fehling's solution 2. Dissolve 12 g of sodium hydroxide pellets and 34 g of sodium potassium 2,3-dihydroxybutanedioate (tartrate), 'Rochelle salt', in water. Filter, if necessary, and make up to 100 cm^3.

Experiment 90. Identifying an unknown carbonyl compound

Requirements per student (or pair)	Notes
A. beaker, 250 cm³	
teat-pipette	
2 test-tubes	
Bunsen burner, tripod, gauze and bench protection mat	
protective gloves	- disposable type preferred.
safety spectacles	
ammonia solution, 2 M NH₃	- ~ 2 cm³. See Experiment 89.
Fehling's solutions, 1 and 2	- ~ 2 cm³ of each. See Experiment 89.
silver nitrate solution, 0.05 M AgNO₃	- ~ 2 cm³. See Experiment 89.
sodium hydroxide solution, 2 M NaOH	- ~ 5 cm³. See Experiment 89.
'unknown' carbonyl compound, X	- ~ 10 cm³. Consult the teacher.
B. beaker, 100 cm³	
beaker, 150 cm³	
beaker, 250 cm³	- or steam-bath.
stirring rod	
suction filtration apparatus	- Büchner or Hirsch funnel, flask, papers, filter pump and tubing.
filter papers	- for drying crystals.
retort stand, boss and clamp	
spatula	
steam bath	- or 250 cm³ beaker.
crushed ice	- on request.
ethanol, C₂H₅OH	- ~ 10 cm³. I.M.S. is suitable.
methanol, CH₃OH	- ~ 2 cm³.
2,4-dinitrophenylhydrazine solution	- ~ 5 cm³. See Experiment 89.
sulphuric acid, dilute, 1 M H₂SO₄	- ~ 2 cm³. See Experiment 89.
C. boiling-tube, fitted with cork, stirrer and thermometer (0-360°C, long-stem)	- See diagram.
3 melting-point tubes	
watch-glass	
rubber ring or band	- thin section of rubber tubing is suitable.
dibutyl phthalate	

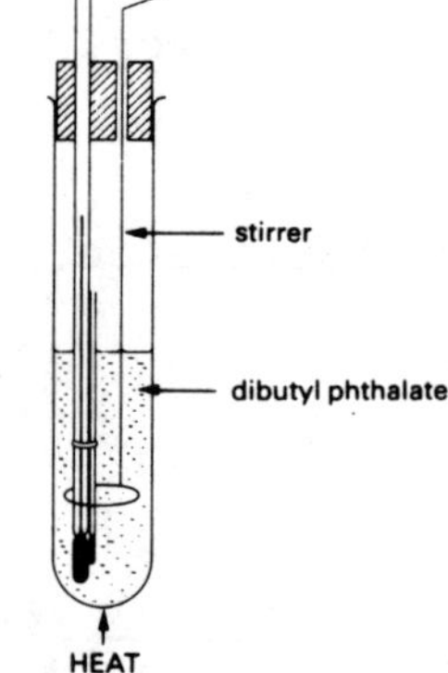

Experiment 91. Chemical properties of carboxylic acids

Requirements per student (or pair)	Notes
teat-pipette	
5 test-tubes	
Bunsen burner and bench mat	
filter papers	- to blot sodium.
forceps or tweezers	- to handle sodium.
spatula	
protective gloves	
safety spectacles	
test-tube rack	
wash-bottle of distilled water	
wood splint	
ethanoic (acetic) acid, glacial, CH_3CO_2H	- ~ 10 cm³. NOT aqueous solution.
phosphorus pentachloride, PCl_5	- ~ 1 g
sodium, metal, Na	- 1 mm cube under the same oil as is used for storage.
sodium ethanoate (acetate), CH_3CO_2Na	- ~ 1 g. The hydrate will serve.
ammonia solution, 2 M NH_3	- ~ 2 cm³. See Experiment 89.
2,4-dinitrophenylhydrazine solution	- ~ 2 cm³. See Experiment 89.
iodine solution, 10% I_2 in KI(aq)	- ~ 2 cm³. See Experiment 89.
iron(III) chloride solution, 0.1 M $FeCl_3$	- ~ 2 cm³. Dissolve 2.5 g of solid in water with 2 cm³ of concentrated hydrochloric acid and make up to 100 cm³.
limewater, saturated, $Ca(OH)_2$	- ~ 2 cm³. See below.
sodium hydrogencarbonate solution, saturated, $NaHCO_3$	- ~ 2 cm³. Shake about 9 g of solid in 100 cm³ of water. Decant from excess solid after standing.
sodium hydroxide solution, 2 M NaOH	- ~ 2 cm³. See Experiment 89.
universal indicator solution	- in dropping bottle.

Limewater. Shake 75 g of calcium hydroxide in 2 dm³ of water in a Winchester
bottle and allow to settle. Siphon off as required and refill with water.

Experiment 92. Identifying salts of carboxylic acids

Requirements per student (or pair)	Notes
delivery tube with bung	- see diagram.

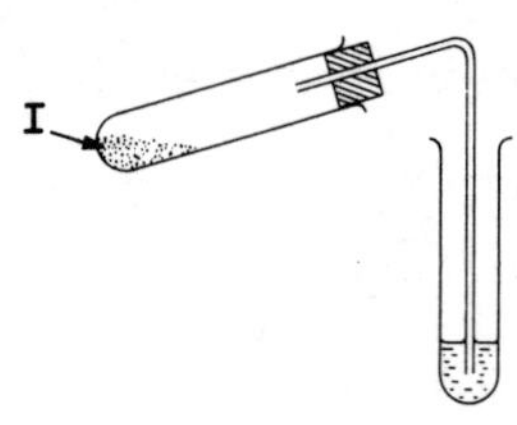

teat-pipette	
6 test-tubes, 2 with corks	
Bunsen burner and bench mat	
litmus papers or pH papers	
protective gloves	
safety spectacles	
spatula	
test-tube holder	
test-tube rack	
wash-bottle of distilled water	
wood splint	
calcium ethanoate, $(CH_3CO_2)Ca$	- ~ 5 g. Labelled 'UNKNOWN I'.
calcium methanoate, $(HCO_2)_2Ca$	- ~ 5 g. Labelled 'UNKNOWN J'.
2,4-dinitrophenylhydrazine solution $C_6H_3(NO_2)_2NHNH_2$	- ~ 2 cm³. See Experiment 89.
limewater, saturated, $Ca(OH)_2$	- issue on request.
sodium hydroxide solution, 2 M NaOH	- ~ 5 cm³. See sheet Experiment 89.
sulphuric acid, concentrated, H_2SO_4	- ~ 2 cm³.
sulphuric acid, dilute, 1 M H_2SO_4	- ~ 2 cm³. See sheet Experiment 89.

Experiment 93. Chemical properties of ethanoyl chloride

Requirements per student (or pair)	Notes
4 beakers, 100 cm^3	
stirring rod	
4 teat-pipettes	
test-tube	
protective gloves	
safety spectacles	
universal indicator paper	
wash-bottle of distilled water	
ethanol, C_2H_5OH	- ~ 5 cm^3. I.M.S. (74 op) is suitable.
ethanoyl (acetyl) chloride, CH_3COCl	- ~ 2 cm^3.
phenylamine (aniline), $C_6H_5NH_2$	- ~ 1 cm^3. Fresh sample - see below.
ammonia solution, 0.880 NH_3	- ~ 5 cm^3.
ammonia solution, 2 M NH_3	- ~ 10 cm^3. See sheet Experiment 89.
iron(III) chloride solution, 0.1 M $FeCl_3$	- ~ 1 cm^3. See sheet Experiment 91.
sodium carbonate solution, 1 M Na_2CO_3	- ~ 1 cm^3. Dissolve 106 g of anhydrous solid in water and make up to 1000 cm^3.

Phenylamine (aniline) is colourless when pure, but darkens on storage due to
aerial oxidation. It is best to purchase small quantities regularly, but
stock that has darkened beyond a clear yellow-brown colour can be purified by
distillation in a fume cupboard. Add a little powdered zinc and distil
using an air condenser, collecting the fraction at 183-185 °C.

Experiment 94. Preparing an ester

Requirements per student (or pair)	Notes
A. conical flask, 250 cm^3, with tight-fitting rubber bung	
measuring cylinder, 10 cm^3	
measuring cylinder, 100 cm^3	
weighing-bottle	
suction filtration apparatus	– Buchner or Hirsch funnel, flask, papers, filter pump and tubing.
protective gloves	
safety spectacles	
spatula	
wash-bottle of distilled water	
access to balance,	– sensitivity ± 0.01 g
benzoyl chloride, C_6H_5COCl	– ~ 10 cm^3.
phenol, C_6H_5OH	– ~ 5 g.
sodium hydroxide solution, 2 M NaOH	– ~ 100 cm^3. See Experiment 89.
B. boiling-tube	
beaker, 250 cm^3	– or water-bath
specimen bottle, with label	
stirring rod	
thermometer, 0-100 °C	
Bunsen burner, tripod, gauze and bench protection mat	
filter papers	– for drying crystals.
water bath (thermostatic control)	– if available, set at 60 °C.
crushed ice	
ethanol, C_2H_5OH	– ~ 25 cm^3. I.M.S. is suitable.
C. boiling-tube fitted with cork, stirrer and thermometer (0-100 °C, long stem)	– see Experiment 90.
3 melting-point tubes	
watch-glass	
retort stand, boss and clamp	
rubber ring or band	– for fixing melting-point tubes to thermometer.
dibutyl phthalate	– ~ 15 cm^3.

Experiment 95. Observation and deduction exercise

Requirements per student (or pair)	Notes
2 beakers, 100 cm³	
beaker, 250 cm³	
boiling-tube	
stirring rod	
teat-pipette	
6 test-tubes	
Bunsen burner, tripod, gauze and mat	
crucible lid	
safety spectacles and gloves	
spatula	
test-tube holder	- preferably metal.
test-tube rack	
wash-bottle of distilled water	
2 wood splints	
4-oxopentanoic (laevulinic) acid* solution, 10% $CH_3COCH_2CH_2CO_2H$	- ~ 10 cm³. Labelled 'UNKNOWN F'. Dissolve 10 g of solid in water and make up to 100 cm³.
benzaldehyde, C_6H_5CHO	- ~ 5 cm³. Labelled 'UNKNOWN G'.
ethanoic (acetic) anhydride, $(CH_3CO)_2O$	- ~ 15 cm³. Labelled 'UNKNOWN H'. Fresh sample - consult teacher.
ethanoic (acetic) acid, glacial, CH_3CO_2H	- ~ 2 cm³. NOT aqueous solution.
phenylamine (aniline), $C_6H_5NH_2$	- ~ 2 cm³. Fresh sample - Experiment 93.
phosphorus pentachloride, PCl_5	- ~ 1 g.
sodium carbonate, Na_2CO_3	- ~ 1 g. Anhydrous.
sodium hydrogencarbonate, $NaHCO_3$	- ~ 1 g.
ammonia solution, 2 M NH_3	- ~ 2 cm³. See Experiment 89.
2-4, dinitrophenylhydrazine solution	- ~ 2 cm³. See Experiment 89.
hydrochloric acid, concentrated, HCl	- ~ 1 cm³.
limewater, saturated, $Ca(OH)_2$	- on request. See sheet Experiment 91.
potassium iodide solution, 0.6 M KI	- ~ 3 cm³. Dissolve 10 g of solid in water and make up to 100 cm³.
silver nitrate solution, 0.05 M $AgNO_3$	- ~ 10 cm³. See sheet Experiment 89.
sodium chlorate(I) (hypochlorite) solution	- ~ 10 cm³. Fresh sample of 10-14% solution. Consult teacher.
sodium hydroxide solution, 2 M NaOH	- ~ 2 cm³. See sheet Experiment 89.
sodium hydroxide solution, 6 M NaOH	- ~ 2 cm³. Dissolve 24 g of pellets in water and make up to 100 cm³.

*This may not be available from your usual supplier. Try Sigma London Chemical Company Ltd, Poole, Dorset BH17 7NH. Order by telephone (FREEFONE SIGMA) - levulinic acid, L 0626.

Experiment 96. The glycine/copper(II) complex

Requirements per student (or pair)	Notes
crystallizing dish	
filter flask or filter tube	- to fit Büchner or Hirsch funnel
stirring rod	
test-tube	
Büchner funnel, small	- a Hirsch funnel may be better, if available.
filter papers	- to fit funnel
filter pump and pressure-tubing	- to fit filter tube or flask
safety spectacles	
spatula	
wash-bottle of distilled water	
copper(II) carbonate, $CuCO_3$	- ~ 3 g. Finely powdered.
glycine (aminoethanoic acid), $CH_2NH_2CO_2H$	- ~ 1 g.

Experiment 97. The biuret test for proteins

Requirements per student (or pair)	Notes
teat-pipette	
4 test-tubes	
safety spectacles	
spatula	
test-tube rack	
wash-bottle of distilled water	
3 protein samples	- ~ 0.5 g of solid in 1 cm^3 of solution, e.g. egg albumin, gelatin, fresh milk.
copper(II) sulphate solution, 0.2 M $CuSO_4$	- ~ 2 cm^3. Dissolve 2.5 g of $CuSO_4 \cdot 5H_2O$ in water and make up to 100 cm^3.
sodium hydroxide solution, 2 M NaOH	- ~ 5 cm^3. Carefully dissolve 80 g of pellets in water and make up to 1000 cm^3.

Experiment 98. Paper chromatography

Requirements per student (or pair)

beaker, 400 cm^3, tall form

measuring cylinder, 10 cm^3

4 melting-point tubes

watch-glass, large

chromatography paper square,
   Whatman no. 1, 12.5 cm x 12.5 cm

fume cupboard access

hair drier

oven

2 paper-clips

pencil and ruler

protective gloves

2 retort stands, bosses and clamps

safety spectacles

wash-bottle of distilled water

ethanol, C_2H_5OH

ammonia solution, 0.880 NH_3

aspartic acid solution, 0.01 M

leucine solution, 0.01 M

lysine solution, 0.01 M

mixture of the three amino acids
   above

ninhydrin solution, aerosol spray

Notes

- chromatography paper must
  fit inside.

- to cover the beaker.

- this paper must not be touched
  with fingers except at the
  edges. Wear clean gloves to
  cut standard 25 cm squares into
  four and store in a box.

- shared by all students

- to be set at 105° - 110°.
  Shared by all students.

- HB or B pencil. 30 cm ruler.

- ~ 25 cm^3. I.M.S. is suitable

- ~ 5 cm^3.

- ~ 1 cm^3. Dissolve 0.13 g of
  solid in 10 cm^3 of propan-2-ol
  and make up to 100 cm^3 with
  water.

- ~ 1 cm^3. Dissolve 0.13 g of
  solid in 10 cm^3 of propan-2-ol
  and make up to 100 cm^3 with
  water.

- ~ 1 cm^3. Dissolve 0.18 g of
  lysine hydrochloride in 10 cm^3
  of propan-2-ol and make up to
  100 cm^3 with water.

- ~ 1 cm^3. Dissolve the masses
  stated above in 10 cm^3 of
  propan-2-ol and make up to
  100 cm^3 with water.

Experiment 99. Reactions of carbohydrates

Requirements per student (or pair)	Notes
General requirements - all parts	
beaker, 250 cm^3	- for use as a water-bath.
6 test-tubes	
Bunsen burner, tripod, gauze and bench mat	
safety spectacles	
spatula	
test-tube rack	
wash-bottle of distilled water	
fructose, $C_6H_{12}O_6$	- $\sim$ 1 g.
glucose, $C_6H_{12}O_6$	- $\sim$ 5 g.
maltose, $C_{12}H_{22}O_{11}$	- $\sim$ 1 g. Not essential.
sucrose, $C_{12}H_{22}O_{11}$	- $\sim$ 50 g. Granulated sugar will do.
starch solution, 2%	- $\sim$ 5 cm^3. See next sheet for details of making up all solutions
sulphuric acid, dilute, 1 M H_2SO_4	- $\sim$ 5 cm^3.
Part A	
protective gloves	
wood splint	
sulphuric acid, concentrated, H_2SO_4	- $\sim$ 2 cm^3.
limewater, saturated, $Ca(OH)_2(aq)$	- $\sim$ 5 cm^3.
potassium dichromate(VI) solution, 0.1 M $K_2Cr_2O_7$	- $\sim$ 5 cm^3.
4 strips of filter paper, 0.5 cm wide	- for testing gases.
Part B	
ammonia solution, 2 M NH_3	- $\sim$ 5 cm^3.
Fehling's solutions, 1 and 2	- $\sim$ 5 cm^3 of each.
silver nitrate solution, 0.05 M $AgNO_3$	- $\sim$ 5 cm^3.
Part C	
phenylhydrazinium chloride, $C_6H_5NHNH_3Cl$	- $\sim$ 1 g.
sodium ethanoate-3-water, $CH_3CO_2Na \cdot 3H_2O$	- $\sim$ 2 g.
2,4-dinitrophenylhydrazine solution	- $\sim$ 10 cm^3.
Part D	
litmus paper	
ammonia solution, 2 M NH_3	- $\sim$ 5 cm^3.
Fehling's solutions, 1 and 2	- $\sim$ 5 cm^3 of each.
iodine solution, 0.1 M I_2 in KI(aq)	- $\sim$ 2 cm^3.
Part E	
polarimeter, simple type.	

Note for Part C: (Students who have time available may also ask for ethanol (IMS), and apparatus for suction filtration and determination of melting-point.)

Experiment 99. (Solutions required.)

<u>Ammonia solution, 2 M NH_3.</u> Carefully dilute 45 cm^3 of 0.880 ammonia to 1000 cm^3 with distilled water.

<u>2,4-dinitrophenylhydrazine solution.</u> Dissolve 0.5 g of solid in 25 cm^3 of methanol and add, dropwise, 1 cm^3 of concentrated sulphuric acid. Filter off any undissolved solid.

<u>Fehling's solution 1.</u> Dissolve 7 g of copper(II) sulphate crystals in water containing a few drops of dilute sulphuric acid, and make up to 100 cm^3.

<u>Fehling's solution 2.</u> Dissolve 12 g of sodium hydroxide pellets and 34 g of sodium potassium 2,3-dihydroxybutanedioate (tartrate), commonly called 'Rochelle salt', in distilled water. Filter if necessary and make up to 100 cm^3.

<u>Iodine solution, 0.1 M I_2 in KI(aq).</u> Dissolve 4 g of potassium iodide, KI, in 20 cm^3 of distilled water, add 2.5 g of iodine and stir to dissolve. Dilute to 100 cm^3.

<u>Limewater, saturated, $Ca(OH)_2$(aq).</u> Shake 75 g of calcium hydroxide in 2 dm^3 of water in a Winchester bottle and allow to settle. Fit the bottle with a soda-lime guard tube and a siphon. Siphon off as required and refill with distilled water.

<u>Potassium dichromate solution, 0.1 M $K_2Cr_2O_7$.</u> Dissolve 2.9 g of solid in distilled water and make up to 100 cm^3.

<u>Silver nitrate solution, 0.05 M $AgNO_3$.</u> Dissolve 2.1 g of solid in distilled water and make up to 250 cm^3. Store and dispense in amber bottles.

<u>Sulphuric acid, dilute, 1 M H_2SO_4.</u> Carefully add 55 cm^3 of concentrated acid, slowly and with constant stirring, to 800 cm^3 of distilled water and make up to 1000 cm^3.

<u>Starch solution, 2%.</u> Mix 2 g of <u>soluble</u> starch to a paste with a little distilled water. Add the paste to 100 cm^3 of boiling distilled water, with stirring, and boil for 1 minute. Cool. Keep for no longer than two weeks.

MASTER SHEETS

for

Advanced Practical Chemistry

Specimen Results and Answers to Questions
Enlarged Results Tables

SPECIMEN RESULTS AND ANSWERS TO QUESTIONS
FOR ADVANCED PRACTICAL CHEMISTRY

The University of London School Examinations Board accepts no responsibility whatsoever for the accuracy or method of working in the answers given for the following experiments: 38, 60, 61, 67, 75, 80, 88, and 95.

Experiment 1. Specimen results and calculations

Results Table 1

Number of drops to deliver 1.0 cm^3 of solution	Number of drops delivered to make monomolecular layer	Diameter of monomolecular layer/cm
100	12	11.2

1. Volume of 1 drop $= \dfrac{1.0 \text{ cm}^3}{100} = \boxed{0.010 \text{ cm}^3}$

2. Since 1000 cm^3 of solution contains 0.050 cm^3 of oleic acid

$$1.00 \text{ cm}^3 \text{ of solution contains } \frac{0.050 \text{ cm}^3}{1000} \text{ of oleic acid}$$

$$\therefore \quad 0.010 \text{ cm}^3 \text{ of solution contains } \frac{0.050 \text{ cm}^3 \times 0.010}{1000} \text{ of oleic acid}$$

$$= \boxed{5.0 \times 10^{-7} \text{ cm}^3}$$

3. Volume of oleic acid in monomolecular layer $= 5.0 \times 10^{-7} \text{ cm}^3 \times 12$

$$= \boxed{6.0 \times 10^{-6} \text{ cm}^3}$$

4. $A = \dfrac{3.142 \times (11.2 \text{ cm})^2}{4} = \boxed{98.5 \text{ cm}^2}$

5. Thickness $= \dfrac{\text{volume}}{\text{area}} = \dfrac{6.0 \times 10^{-6} \text{ cm}^3}{98.5 \text{ cm}^2} = \boxed{6.1 \times 10^{-8} \text{ cm}}$

6. Volume of one molecule $= (6.1 \times 10^{-8} \text{ cm})^3 = \boxed{2.3 \times 10^{-22} \text{ cm}^3}$

7. Density $= \dfrac{\text{molar mass}}{\text{molar volume}}$

$$\therefore \text{ molar volume} = \frac{\text{molar mass}}{\text{density}} = \frac{282 \text{ g mol}^{-1}}{0.890 \text{ g cm}^{-3}} = \boxed{317 \text{ cm}^3 \text{ mol}^{-1}}$$

8. $L = \dfrac{\text{molar volume}}{\text{volume of molecule}} = \dfrac{317 \text{ cm}^3 \text{ mol}^{-1}}{2.3 \times 10^{-22} \text{ cm}^3} = \boxed{1.4 \times 10^{24} \text{ mol}^{-1}}$

(The accepted value of L is 6.02×10^{23} mol^{-1}. You should expect to obtain a value of L to within one power of 10, i.e. between 6×10^{22} mol^{-1} and 6×10^{24} mol^{-1}.)

Experiment 1. Questions

1. The various sources of error which may account for the difference between
    the experimental and actual value for L are as follows:

    (a) the number of drops required to make a monomolecular layer was
        inaccurately measured;

    (b) not all the pentane had evaporated from the surface;

    (c) the volume of the drop was inaccurately determined;

    (d) the molecules were unevenly dispersed on the surface of the film,
        i.e. there was more than one layer of molecules on parts of the film;

    (e) the assumption about the shape of the molecule was incorrect. In
        fact, the molecule is rather more the shape of a cylinder than a
        cube.

2. The number of drops required to fill the loop is subject to the greatest
    error. Even if you are sure to the nearest drop, this gives only two
    significant figures (or only one for a small loop!) whereas all the
    other values are obtained to three significant figures.

3. A substitute for pentane must:

    (a) dissolve oleic acid readily;

    (b) evaporate readily;

    (c) not react with oleic acid;

    (d) not react with water;

    (e) not dissolve in water.

Experiment 2. Specimen results

Results Table 2

Molar mass of potassium hydrogenphthalate, M	204.1 g mol^{-1}
Mass of bottle and contents before transfer, m_1	15.47 g
Mass of bottle and contents after transfer, m_2	10.20 g
Mass of potassium hydrogenphthalate, $m = (m_1 - m_2)$	5.27 g
Amount of potassium hydrogenphthalate, $n = m/M$	2.58×10^{-2} mol
Volume of solution, V	0.250 dm^3
Concentration of potassium hydrogenphthalate, $c = n/V$	0.103 mol dm^{-3}

Experiment 2. Questions

1. (a) The concentration would be lower than calculated. From the expression
       $c = n/V$, a decrease in amount, n, will reduce the value of the
       concentration.

   (b) The concentration would be greater than calculated. From the expres-
       sion $c = n/V$, insufficient water (i.e. a decrease in V) will increase
       the value of the concentration.

Experiment 3. Specimen results

Pipette solution	potassium hydrogenphthalate		0.0103	mol dm^{-3}	25.0 cm^3
Burette solution	sodium hydroxide		?	mol dm^{-3}	
Indicator	phenolphthalein				
		Trial	1	2	3	(4)
Burette readings	Final	26.9	26.80	27.45	26.15	-
	Initial	0.7	0.90	1.50	0.30	-
Volume used (titre)/cm^3		26.2	25.90	25.95	25.85	-
Mean titre/cm^3		25.9(0)				

Calculation

Let A refer to potassium hydrogenphthalate and B to sodium hydroxide.

1. Substituting into the expression

$$\frac{c_A V_A}{c_B V_B} = \frac{a}{b}$$

where c_A = 0.103 mol dm^{-3} c_B = ?

V_A = 25.0 cm^3 V_B = 25.9 cm^3

a = 1 b = 1

gives $\dfrac{0.103 \text{ mol dm}^{-3} \times 25.0 \text{ cm}^3}{c_B \times 25.9 \text{ cm}^3} = \dfrac{1}{1}$

$\therefore c_B = \dfrac{0.103 \text{ mol dm}^{-3} \times 25.0 \text{ cm}^3}{25.9 \text{ cm}^3} = \boxed{0.0994 \text{ mol dm}^{-3}}$

Experiment 3. Questions

1. (a) If the burette is dry, no effect. If the burette is wet, this will slightly dilute the NaOH so that the calculated concentration will be less than the actual concentration.

    (b) If the pipette is dry, no effect. If the pipette is wet, this will slightly dilute the measured volume of HA so that less NaOH will be required to neutralise it. The calculated concentration of NaOH will therefore be lower than the actual concentration.

    (c) The calculated concentration of NaOH will be lower than the actual concentration.

    (d) No effect. The dilution occurs after the amount has been measured.

2. It is easier for the eye to detect the approach of the end-point when colour appears rather than when it disappears.

3. Sodium hydroxide solution attacks glass to some extent, especially the ground surfaces in some burette taps. Also, it absorbs carbon dioxide from the air forming a crust of sodium carbonate. Both these actions can cause burette taps to seize up.

Experiment 4. Specimen results

Results Table 4

Pipette solution	iodine				0.0497 mol dm^{-3}	10.0 cm^3
Burette solution	sodium thiosulphate				0.0512 mol dm^{-3}	
Indicator	starch					
		Trial	1	2	3	(4)
Burette readings	Final	20.3	40.40	20.15	40.25	–
	Initial	0.0	20.30	0.10	20.15	–
Volume used (titre)/cm^3		20.3	20.10	20.05	20.10	–
Mean titre/cm^3		20.1(0)				

<u>Calculation</u>

1. Substituting into the expression

$$\frac{c_A V_A}{c_B V_B} = \frac{a}{b} \qquad \text{(A refers to Na}_2\text{S}_2\text{O}_3\text{, B to I}_2\text{)}$$

where $c_A = 0.0512$ mol dm^{-3} $\qquad c_B = 0.0497$ mol dm^{-3}

$\qquad V_A = 20.1$ cm^3 $\qquad\qquad V_B = 10.0$ cm^3

$\qquad a = ?$ $\qquad\qquad\qquad b = ?$

gives $\dfrac{a}{b} = \dfrac{0.0512 \text{ mol dm}^{-3} \times 20.1 \text{ cm}^3}{0.0497 \text{ mol dm}^{-3} \times 10.0 \text{ cm}^3} = \dfrac{2.07}{1}$

$\therefore a = 2$ and $b = 1$

So we can write for the equation:

$$2\text{Na}_2\text{S}_2\text{O}_3(\text{aq}) + \text{I}_2(\text{aq}) \rightarrow \text{Products}$$

2. The formula for the other compound is $\text{Na}_2\text{S}_4\text{O}_6$ (sodium tetrathionate).

$$2\text{Na}_2\text{S}_2\text{O}_3(\text{aq}) + \text{I}_2(\text{aq}) \rightarrow \text{NaI}(\text{aq}) + ?$$

To balance I atoms, the stoichiometric coefficient of NaI must be 2.

$$2\text{Na}_2\text{S}_2\text{O}_3(\text{aq}) + \text{I}_2(\text{aq}) \rightarrow 2\text{NaI} + ?$$

The atoms unaccounted for are two of sodium, four of sulphur and six of oxygen, $\text{Na}_2\text{S}_4\text{O}_6$.

The balanced equation is

$$\boxed{2\text{Na}_2\text{S}_2\text{O}_3(\text{aq}) + \text{I}_2(\text{aq}) \rightarrow 2\text{NaI}(\text{aq}) + \text{Na}_2\text{S}_4\text{O}_6(\text{aq})}$$

(The empirical formula is NaS_2O_3, but $\text{Na}_2\text{S}_4\text{O}_6$ is preferred because it contains $\text{S}_4\text{O}_6{}^{2-}$ ions.)

Experiment 5. Specimen results

Results Table 5a

Mass of bottle and contents before transfer, m_1	11.79 g
Mass of bottle and contents after transfer, m_2	10.21 g
Mass of sample, $m = (m_1 - m_2)$	1.58 g
Mass of $BaCl_2 \cdot xH_2O$ in 10.0 cm^3	0.0632 g

Results Table 5b

Pipette solution	barium chloride		?	mol dm^{-3}	10.0 cm^3
Burette solution	silver nitrate		0.0506	mol dm^{-3}	
Indicator	potassium chromate(VI)				
		Trial 1	2	3	(4)
Burette readings	Final	10.4	20.65	30.40	41.10	–
	Initial	0.0	10.40	20.75	30.90	–
Volume used (titre)/cm^3		10.4	10.25	10.15	10.20	–
Mean titre/cm^3		10.2(0)			

Experiment 5. Calculation

There is more than one way of tackling this problem. Here is one suggestion.

First calculate the masses of $BaCl_2$ and H_2O in each sample. The mass of $BaCl_2$ can be calculated indirectly from titration data and the mass of H_2O by difference.

Next convert masses to relative amounts to determine x. We explain this step by step.

1. Amount of Cl in sample (as Cl^-)

$$Ag^+(aq) + Cl^-(aq) \rightarrow AgCl(s)$$

$\therefore$ amount of Cl = amount of $AgNO_3$

$$= cV = 0.0506 \text{ mol dm}^{-3} \times 0.0102 \text{ dm}^3 = 5.16 \times 10^{-4} \text{ mol}$$

2. Amount and mass of $BaCl_2$

$$BaCl_2(s) + aq \rightarrow Ba^{2+}(aq) + 2Cl^-(aq)$$

$\therefore$ amount of $BaCl_2 = \frac{1}{2} \times$ amount of Cl^-

$$= \frac{1}{2} \times 5.16 \times 10^{-4} \text{ mol} = 2.58 \times 10^{-4} \text{ mol}$$

mass of $BaCl_2$ $= nM = 2.58 \times 10^{-4} \text{ mol} \times 208 \text{ g mol}^{-1} = 0.0537 \text{ g}$

(Continued on next page.)

(Continued from last page.)

3. __Mass and amount of H_2O__

$$\text{mass of } H_2O = \text{mass of } BaCl_2 \cdot xH_2O - \text{mass of } BaCl_2$$

$$= 0.0632 \text{ g} - 0.0537 \text{ g} = 0.0095 \text{ g}$$

$$\text{amount of } H_2O = \frac{m}{M} = \frac{0.0095 \text{ g}}{18.0 \text{ g mol}^{-1}} = 5.3 \times 10^{-4} \text{ mol}$$

4. __Relative amounts__

	$BaCl_2$	H_2O
Amount/mol	2.58×10^{-4}	5.3×10^{-4}
Amount/smallest amount	$\dfrac{2.58 \times 10^{-4}}{2.58 \times 10^{-4}}$	$\dfrac{5.3 \times 10^{-4}}{2.58 \times 10^{-4}}$
= relative amount	= 1.00	= 2.1

The relative amounts are very close to the integers 1 and 2.

$\therefore$ $\boxed{x = 2 \text{ and the formula is } BaCl_2 \cdot 2H_2O}$

Experiment 6. Specimen results

Results Table 6

Pipette solution	Ammonium iron(II) sulphate		39.2 g dm^{-3}		25.0 cm^3
Burette solution	Potassium manganate(VII)		$0.0200 \text{ mol dm}^{-3}$		
Indicator	Self-indicating				
		Trial	1	2	3	(4)
Burette readings	Final	25.7	25.10	25.05	25.15	
	Initial	0.5	0.00	0.00	0.00	
Volume/used (titre) cm³			25.2	25.10	25.05	25.15	
Mean titre/cm³			25.1				

(Continued on next page.)

(Continued from last page.)

1. Substituting into the expression

$$\frac{c_A V_A}{c_B V_B} = \frac{a}{b}$$

(A refers to MnO_4^-, B to Fe^{2+})

where c_A = 0.0200 mol dm^{-3} $\qquad$ c_B = ?

$\qquad\quad$ V_A = 25.1 cm^3 $\qquad\qquad$ V_B = 25.0 cm^3

$\qquad\quad$ a = 1 $\qquad\qquad\qquad$ b = 5

gives $\quad \dfrac{0.0200 \text{ mol dm}^{-3} \times 25.1 \text{ cm}^3}{c_B \times 25.0 \text{ cm}^3} = \dfrac{1}{5}$

$\therefore \quad c_B = \dfrac{0.0200 \text{ mol dm}^{-3} \times 25.1 \text{ cm}^3 \times 5}{25.0 \text{ cm}^3} = 0.100 \text{ mol dm}^3$

2. Mass of $Fe(NH_4)_2(SO_4)_2$ in one litre

    = amount x molar mass

    = 0.100 mol x 284 g mol^{-1} = 28.4 g

3. Mass of water in 39.2 g of salt = (39.2 - 28.4) g = 10.8 g

4. Since 39.2 g is 0.100 mol of the salt, the amount of water in one mole
    of the salt is given by

$$n = \frac{1.08 \text{ g}}{18 \text{ g mol}^{-1}} \times \frac{1}{0.100} = 6 \text{ mol}$$

$\therefore \quad \boxed{x = 6 \text{ and the formula is } Fe(NH_4)_2(SO_4)_2 \cdot 6H_2O}$

Experiment 7. Questions

(a) A current flows between the grid and the cathode before ionization
     because the electrons repelled by the negative valve anode return to
     the grid and then flow round the circuit.

(b) Every argon atom which ionizes produces an extra electron which
     returns to the grid. This causes a sudden increase in grid current.

Experiment 8. Questions

A continuous spectrum shows a continuous range of wavelengths. In the visible
range, this appears as broad bands of colour merging into one another with no
sharp boundaries.

An emission spectrum shows a limited number of wavelengths, which appear as
distinct lines. In the visible range, the lines appear in different colours.

Experiment 9. Specimen results and calculations

Results Table 9

Time/min	0.0	0.5	1.0	1.5	2.0	2.5	3.0	3.5	4.0	4.5
Temperature/$^{\circ}$C	27.0	27.0	27.2	27.2	27.2	27.2	–	66.0	71.4	71.8
Time/min	5.0	5.5	6.0	6.5	7.0	7.5	8.0	8.5	9.0	9.5
Temperature/$^{\circ}$C	70.2	68.0	66.2	64.4	62.8	61.0	59.5	58.0	56.6	55.1

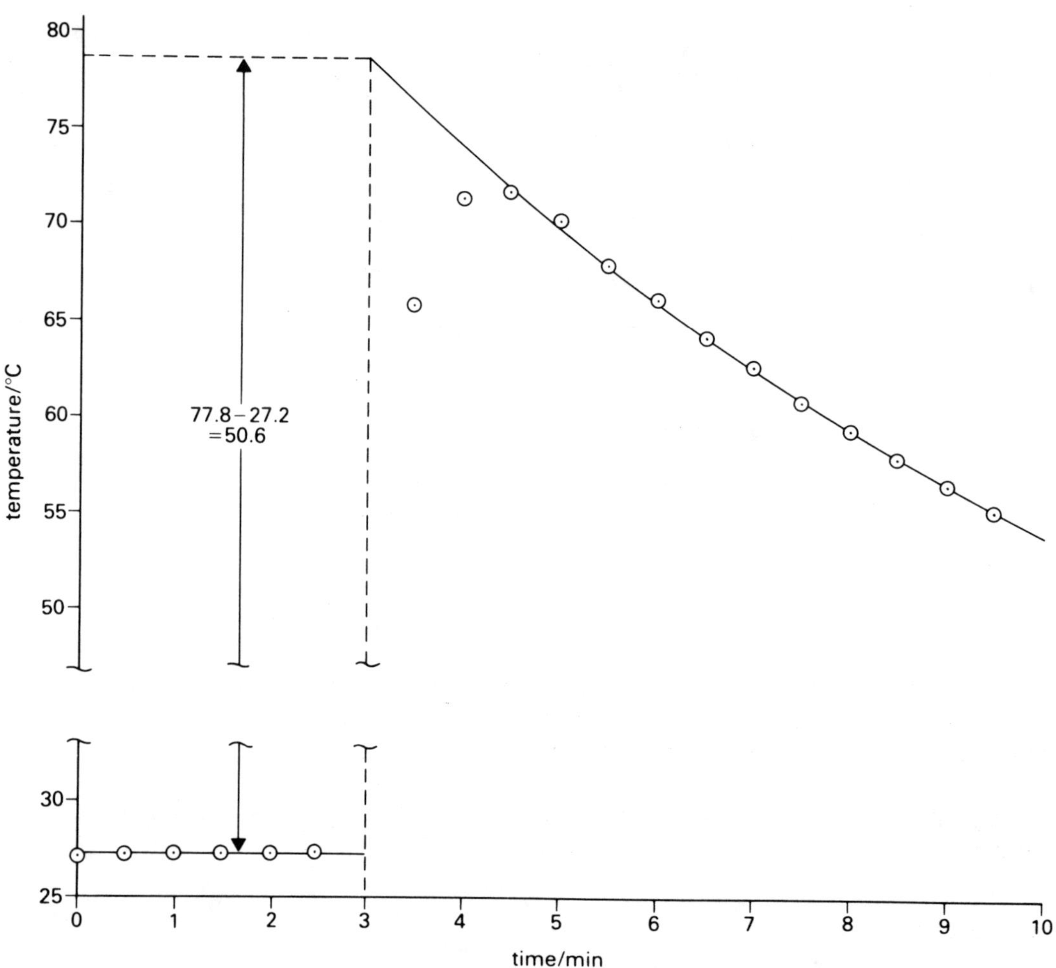

$$\left[\begin{array}{l}\text{enthalpy change due to}\\ \text{reaction (at constant } T)\end{array}\right] + \left[\begin{array}{l}\text{change in heat}\\ \text{energy of solution}\end{array}\right] = 0$$

i.e. $\qquad \Delta H \qquad + \qquad mc_p\Delta T \qquad = 0$

$\therefore \Delta H = -mc_p\Delta T = -0.0250$ kg $\times$ 4.18 kJ kg^{-1} K^{-1} $\times$ 50.6 K $=$ $\boxed{-5.29 \text{ kJ}}$

Amount of Cu^{2+} used $= cV = 1.00$ mol dm^{-3} $\times$ 0.0250 dm^{-3} $=$ 0.0250 mol

Scaling up to the amount in the equation, 1 mol,

$$\Delta H = -5.29 \text{ kJ} \times \frac{1}{0.0250} = \boxed{-212 \text{ kJ}}$$

$$Zn(s) + Cu^{2+}(aq) \rightarrow Cu(s) + Zn^{2+}(aq); \quad \Delta H^{\ominus} = \boxed{-212 \text{ kJ mol}^{-1}}$$

Experiment 9. Questions

1. $\text{Error} = \dfrac{(-212 - -217) \text{ kJ mol}^{-1}}{-217 \text{ kJ mol}^{-1}} \times 100 = \dfrac{+5}{-217} \times 100 = -2.3\%$

2. (a) There is some heat loss from the polystyrene cup.
    (b) The heat capacity of the solution is not precisely 4.18 kJ kg^{-1} K^{-1}.
    (c) The density of the solution is not precisely 1.00 g dm^{-3}.
    (d) The heat capacities of the metals were ignored.
    (e) The thermometer has a small but measurable heat capacity.

3. The reaction takes a few minutes to complete because the copper first
    precipitated shields some of each zinc particle from the Cu^{2+}.

Experiment 10. Specimen results and calculations

Suitable amounts are 0.0500 mol of NH_4Cl and 5.00 mol of H_2O.

Then mass of NH_4Cl = 53.4 g mol^{-1} $\times$ 0.500 mol = 2.67 g

and mass of H_2O = 18.0 g mol^{-1} $\times$ 5.00 mol = 90.0 g

ΔT = final temperature - initial temperature = 22.7 °C - 24.5 °C = -1.8 K

$$\left[\begin{array}{l}\text{enthalpy change on}\\ \text{dissolving } NH_4Cl\end{array}\right] + \left[\begin{array}{l}\text{change in heat}\\ \text{energy of solution}\end{array}\right] = 0$$

$$\Delta H \qquad + \qquad mc_p\Delta T \qquad\qquad = 0$$

$\therefore \Delta H = -mc_p\Delta T = -(0.090 \text{ kg} \times 4.18 \text{ kJ kg}^{-1} \text{ K}^{-1} \times (-1.8 \text{ K})) = +0.68 \text{ kJ}$

Scaling up from 0.050 mol to 1 mol,

$\Delta H = +0.68 \text{ kJ} \times \dfrac{1}{0.050} = +14 \text{ kJ}$

$NH_4Cl(s) + 100H_2O(l) \rightarrow NH_4Cl(aq,100H_2O); \quad \Delta H^{\ominus} = \boxed{+14 \text{ kJ mol}^{-1}}$

Note that only two significant figures are justified here because ΔT is small.

Experiment 10. Questions

Results are usually lower than 16.4 kJ mol^{-1}, largely because some heat energy
enters the system to increase the final temperature a little. It is difficult
to measure a small temperature change accurately and an error of only 0.1 K
changes the result by almost 1 kJ mol^{-1}. Also, we have made the usual
assumptions about heat capacities of the polystyrene cup (= 0), thermometer (= 0)
and solution (= water).

Experiment 11. Specimen results and calculations

Results Table 11

	$MgSO_4$	$MgSO_4 \cdot 7H_2O$
Mass of weighing bottle	12.91 g	13.42 g
Mass of weighing bottle and salt	15.92 g	19.58 g
Mass of salt	3.01 g	6.16 g
Mass of polystyrene cup	2.10 g	2.36 g
Mass of polystyrene cup and water	47.10 g	44.21 g
Mass of water	45.00 g	41.85 g
Initial temperature	24.1 °C	24.8 °C
Final temperature	35.4 °C	23.4 °C

(Continued on next page.)

(Continued from last page.)

1. $\Delta H = -mc_p\Delta T = -(0.0450 \text{ kg} \times 4.18 \text{ kJ kg}^{-1} \text{ K}^{-1} \times 11.3 \text{ K}) = -2.12 \text{ kJ}$

Scaling up, $\Delta H = -2.12 \text{ kJ} \times \dfrac{1 \text{ mol}}{0.0250 \text{ mol}} = -84.8 \text{ kJ}$

Thus $MgSO_4(s) + 100H_2O(l) \rightarrow MgSO_4(aq, 100H_2O);\quad \Delta H^\ominus = \boxed{-84.8 \text{ kJ mol}^{-1}}$

2. $\Delta H = -mc_p\Delta T = -(0.04185 \text{ kg} \times 4.18 \text{ kJ kg}^{-1} \text{ K}^{-1} \times (-1.4 \text{ K}) = +0.24 \text{ kJ}$

Scaling up, $\Delta H = +0.24 \text{ kJ} \times \dfrac{1 \text{ mol}}{0.0250 \text{ mol}} = +9.6 \text{ kJ}$

Thus $MgSO_4 \cdot 7H_2O(s) + 93H_2O(l) \rightarrow MgSO_4(aq, 100H_2O);\quad \Delta H^\ominus = \boxed{+9.6 \text{ kJ mol}^{-1}}$

3.

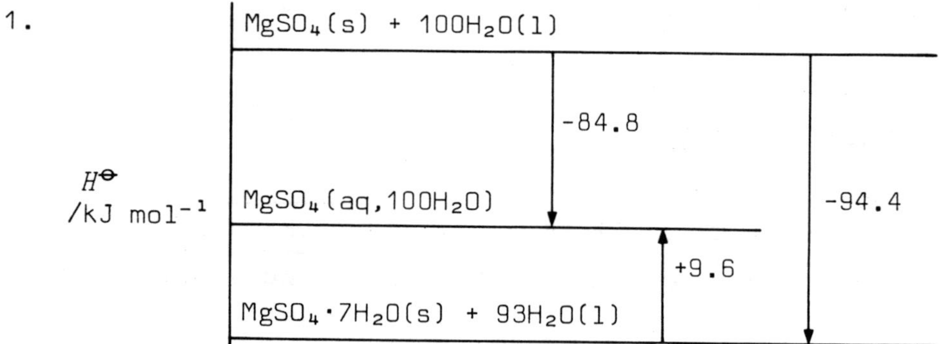

$\Delta H^\ominus = \Delta H_1 - \Delta H_2 = (-84.8 - 9.6) \text{ kJ mol}^{-1} = \boxed{-94.4 \text{ kJ mol}^{-1}}$

Experiment 11. Questions

1.

2. The maximum temperature change is much smaller and occurs more quickly.

3. Error $= \dfrac{-94.4 - (-104.0)}{104.0} \times 100 = -9.2\%$

In addition to the usual assumptions about zero heat transfer between system and surroundings, and the heat capacities of the solutions, impurities in the salts may also affect the results. The anhydrous salt may not be totally free of water, and the hydrated salt may have absorbed extra water.

Experiment 12. Specimen results and calculations

Results Table 12a

Mass of cold water in vacuum flask	50.0 g	50.0 g
Mass of warm water added	50.0 g	50.0 g
Initial temperature of flask and cold water	23.1 °C	23.5 °C
Initial temperature of warm water	41.3 °C	45.2 °C
Final temperature of flask and mixture	31.0 °C	33.0 °C

(Continued on next page.)

(Continued from last page.)

$$\left[\begin{array}{c}\text{Change in heat}\\\text{energy of flask}\end{array}\right] + \left[\begin{array}{c}\text{Change in heat energy}\\\text{of cold water}\end{array}\right] + \left[\begin{array}{c}\text{Change in heat energy}\\\text{of warm water}\end{array}\right] = 0$$

(1) $\quad C \times (31.0 - 23.1)K + 0.0500 \text{ kg} \times 4.18 \text{ kJ kg}^{-1} \text{ K}^{-1} \times (31.0 - 23.1)K$
$$+ 0.0500 \text{ kg} \times 4.18 \text{ kJ kg}^{-1} \text{ K}^{-1} \times (41.3 - 31.0)K = 0$$

$\quad (C \times 7.9 \text{ K}) + (1.65 \text{ kJ}) + (-2.15 \text{ kJ}) = 0$

$\quad \therefore C = \dfrac{(2.15 - 1.65) \text{ kJ}}{7.9 \text{ K}} = 0.063 \text{ kJ K}^{-1}$

(2) $\quad (C \times 9.5 \text{ K}) + (1.99 \text{ kJ}) + (-2.55 \text{ kJ}) = 0$

$\quad \therefore C = \dfrac{(2.55 - 1.99) \text{ kJ}}{9.5 \text{ K}} = 0.059 \text{ kJ K}^{-1}$

$$\text{Mean value of } C = \boxed{0.061 \text{ kJ K}^{-1}}$$

Results Table 12b

Mass of anhydrous copper(II) sulphate (0.025 mol)	3.99 g
Mass of water (0.025 x 100 mol)	45.0 g
Initial temperature of vacuum flask and water	22.5 °C
Maximum temperature of vacuum flask and water	27.5 °C

$$\left[\begin{array}{c}\text{Change in heat}\\\text{energy of flask}\end{array}\right] + \left[\begin{array}{c}\text{Change in heat}\\\text{energy of contents}\end{array}\right] + \left[\begin{array}{c}\text{Enthalpy change}\\\text{of solution}\end{array}\right] = 0$$

$(0.061 \text{ kJ K}^{-1} \times 5.0 \text{ K}) + (0.0450 \text{ kg} \times 4.18 \text{ kJ kg}^{-1} \text{ K}^{-1} \times 5.0 \text{ K}) + \Delta H = 0$

$0.305 \text{ kJ} + 0.941 \text{ kJ} + \Delta H = 0 \qquad \therefore \Delta H = -1.25 \text{ kJ}$

Scaling up to 1 mol, $\Delta H = -1.25 \text{ kJ} \times \dfrac{1}{0.025} = -50 \text{ kJ}$

$\therefore CuSO_4(s) + 100H_2O(l) \rightarrow CuSO_4(aq,100H_2O); \quad \Delta H^{\ominus} = \boxed{-50 \text{ kJ mol}^{-1}}$

Results Table 12c

Mass of copper(II) sulphate-5-water (0.0250 mol)	6.24 g
Mass of water (0.025 x 95 mol)	42.75 g
Initial temperature of vacuum flask and water	23.0 °C
Minimum temperature of vacuum flask and solution	21.8 °C

$$\left[\begin{array}{c}\text{Change in heat}\\\text{energy of flask}\end{array}\right] + \left[\begin{array}{c}\text{Change in heat}\\\text{energy of contents}\end{array}\right] + \left[\begin{array}{c}\text{Enthalpy change}\\\text{of solution}\end{array}\right] = 0$$

$0.061 \text{ kJ K}^{-1} \times (-1.2 \text{ K}) + \left(0.04275 \text{ kg} \times 4.18 \text{ kJ kg}^{-1} \text{ K}^{-1} \times (-1.2 \text{ K})\right) + \Delta H = 0$

$-0.0732 \text{ kJ} + (-0.214 \text{ kJ}) + \Delta H = 0 \qquad \therefore \Delta H = +0.287 \text{ kJ}$

Scaling up to 1 mol, $\Delta H = +0.287 \text{ kJ} \times \dfrac{1}{0.025} = +11 \text{ kJ}$

$\therefore CuSO_4 \cdot 5H_2O(s) + 95H_2O(l) \rightarrow CuSO_4(aq,100H_2O); \quad \Delta H^{\ominus} = \boxed{+11 \text{ kJ mol}^{-1}}$

$$\boxed{95H_2O(l)} + \boxed{CuSO_4(s) + 5H_2O(l)} \xrightarrow{\Delta H^{\ominus}} \boxed{CuSO_4 \cdot 5H_2O(s)} + \boxed{95H_2O(l)}$$

$$\Delta H_1 \searrow \qquad \swarrow \Delta H_2$$

$$\boxed{CuSO_4(aq,100H_2O)}$$

$\Delta H^{\ominus} = \Delta H_1 - \Delta H_2 = -50 \text{ kJ mol}^{-1} - 11 \text{ kJ mol}^{-1} = \boxed{-61 \text{ kJ mol}^{-1}}$

Experiment 12. Questions

1. The reaction would be extremely slow (even with excess water as in the experiment you have done, it was quite slow).

2. Replacing lids and stoppers prevents moisture being absorbed from the air (especially important for anhydrous copper sulphate). Also, stoppers left off can be exchanged by mistake causing contamination of the contents when eventually replaced.

3. It is impossible to transfer <u>all</u> the weighed quantity from a beaker to the flask.

Experiment 13. Specimen results and calculations

Results Table 13a

	1st run	2nd run	
Molar mass of propan-1-ol, M	60.1	60.1	g mol^{-1}
Initial mass of spirit lamp + alcohol, m_1	13.112	12.083	g
Final mass of spirit lamp + alcohol, m_2	12.083	11.034	g
Mass of alcohol burned, $m_1 - m_2$	1.029	1.049	g
Amounts of alcohol burned, $n = (m_1 - m_2)/M$	0.0171	0.0175	mol
Initial temperature of calorimeter	21.7	21.4	°C
Final temperature of calorimeter	33.2	33.0	°C
Temperature change, ΔT	11.5	11.6	K
Heat released during the experiment $= \Delta H_c^{\ominus}$ [propan-1-ol] × amount burned, n	34.5	35.3	kJ
Heat required to raise temperature by 1 K $= \dfrac{-2017 \text{ kJ} \times n}{\Delta T}$ = calibration factor, C	-3.00	-3.04	kJ K^{-1}
Mean value of C		-3.02	kJ K^{-1}

Results Table 13b

	C_4H_9OH	$C_5H_{11}OH$	$C_6H_{13}OH$	$C_7H_{15}OH$	$C_8H_{17}OH$
Molar Mass m/g mol^{-1}	74.1	88.2	102.2	116.2	130.2
Initial mass of lamp/g	13.691	12.820	13.571	13.679	13.909
Final mass of lamp/g	12.642	12.023	12.668	12.794	13.175
Mass of alcohol burned/g	1.049	0.797	0.903	0.885	0.734
Amount burned, n/mol	0.0142	0.00904	0.00884	0.00762	0.00564
Initial temperature/°C	21.5	21.0	21.7	21.4	22.0
Final temperature/°C	33.5	30.0	33.1	32.2	32.1
Temperature change, ΔT/K	12.0	9.0	11.4	10.8	10.1
$\Delta H_c^{\ominus} = \dfrac{C \times \Delta T}{n}$ /kJ mol^{-1}	-2552	-3007	-3895	-4280	-5408

Experiment 13. Questions

1 & 2. See Table 13c

Table 13c

Pair of alcohols	Difference in $\Delta H_C^{\ominus}/\text{kJ mol}^{-1}$	
	Experiment	Data book
Propan-1-ol/butan-1-ol	535	658
Butan-1-ol/pentan-1-ol	455	648
Pentan-1-ol/hexan-1-ol	888	653
Hexan-1-ol/heptan-1-ol	385	647
Heptan-1-ol/octan-1-ol	1128	657
Average difference	678	653

3. The enthalpy of combustion of any gaseous alcohol can be regarded as the
    sum of the enthalpy changes associated with breaking all the bonds in the
    reactants and with forming all the bonds in the products, e.g. for ethanol
    and propanol:

$$
\begin{array}{l}
\text{H} \quad \text{H} \\
| \quad\quad | \\
\text{H—C—C—H} \quad + \quad 3(\text{O}{=}\text{O}) \quad \rightarrow \quad 2(\text{O}{=}\text{C}{=}\text{O}) + 3(\text{H—O—H}) \\
| \quad\quad | \\
\text{H} \quad \text{O—H}
\end{array}
$$

$$\Delta H_C^{\ominus} = 4\bar{E}(\text{C}{=}\text{O}) + 6\bar{E}(\text{O—H}) - \bar{E}(\text{C—C}) - \bar{E}(\text{C—O}) - \bar{E}(\text{O—H}) - 3\bar{E}(\text{O}{=}\text{O}) - 5\bar{E}(\text{C—H})$$

$$
\begin{array}{l}
\text{H} \quad \text{H} \quad \text{H} \\
| \quad\quad | \quad\quad | \\
\text{H—C—C—C—H} + 4\tfrac{1}{2}(\text{O}{=}\text{O}) \quad \rightarrow \quad 3(\text{O}{=}\text{C}{=}\text{O}) + 4(\text{O—O—H}) \\
| \quad\quad | \quad\quad | \\
\text{H} \quad \text{H} \quad \text{O—H}
\end{array}
$$

$$\Delta H_C^{\ominus} = 6\bar{E}(\text{C}{=}\text{O}) + 8\bar{E}(\text{O—H}) - 2\bar{E}(\text{C—C}) - \bar{E}(\text{C—O}) - \bar{E}(\text{O—H}) - 4\tfrac{1}{2}\bar{E}(\text{O}{=}\text{O}) - 7\bar{E}(\text{C—H})$$

The difference between the two enthalpies of combustion is, in effect, due
to the combustion of a $-\text{CH}_2-$ group and is made up as follows:

$$\text{Difference} = 2\bar{E}(\text{C}{=}\text{O}) + 2\bar{E}(\text{O—H}) - \bar{E}(\text{C—C}) - 2\bar{E}(\text{C—H}) - 1\tfrac{1}{2}\bar{E}(\text{O}{=}\text{O})$$

The same expression applies to any adjacent pair of alcohols in the
homologous series.

4. Since the bond energy terms used in the expression for the difference in
    enthalpies of combustion are assumed to be fairly constant, it follows that
    the value for the difference is also fairly constant.

5. Small differences arise for two reasons.

    (a) The C—C bonds and the C—H bonds in adjacent alcohols are not precisely
         identical in that their environments differ slightly. The bond energies
         will therefore differ very slightly whereas the expression derived in
         step 3 assumes constant values.

    (b) A further energy term should, strictly, be included in the calculation,
         namely the enthalpy of vaporization for each alcohol. There is a small
         variation from one alcohol to the next and this introduces a small
         variation in the difference in enthalpy of combustion.

Experiment 14. Specimen results and calculations

Results Table 14a. Titration of hydrochloric acid

Volume added/cm³	0.0	5.0	10.0	15.0	20.0	25.0	30.0	35.0	40.0	45.0	50.0
Temperature/°C	22.2	24.4	26.4	28.4	30.1	31.1	30.4	29.9	29.2	28.8	28.2

Results Table 14b. Titration of ethanoic acid

Volume added/cm³	0.0	5.0	10.0	15.0	20.0	25.0	30.0	35.0	40.0	45.0	50.0
Temperature/°C	21.0	22.6	24.3	25.9	27.2	28.7	28.6	28.0	27.5	27.0	26.5

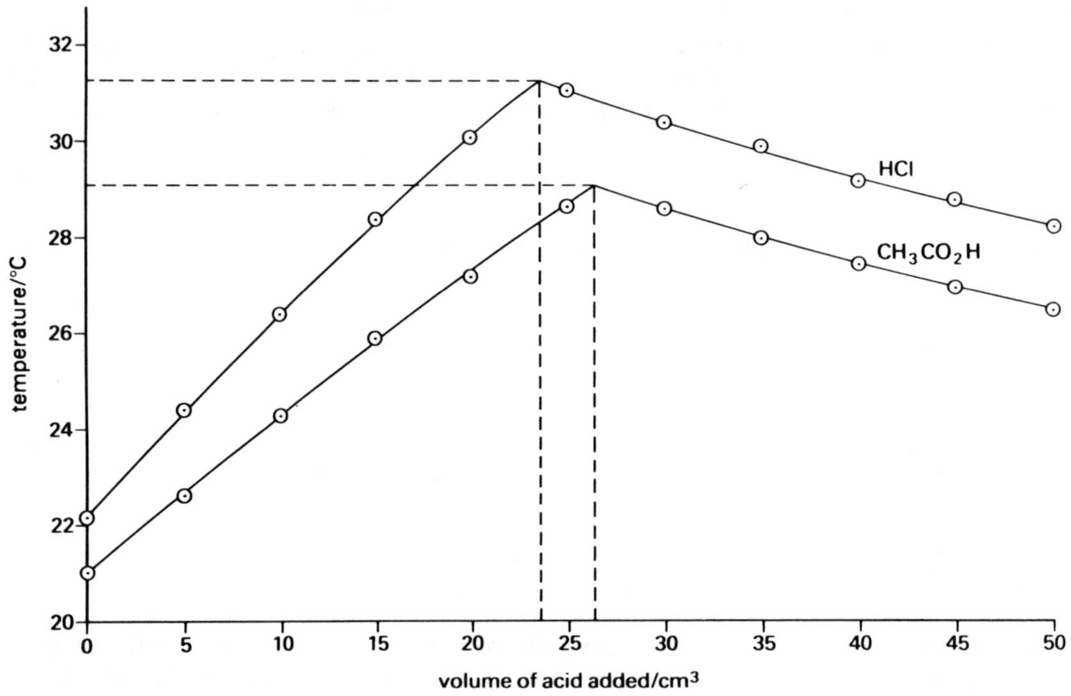

3. (a) $HCl(aq) + NaOH(aq) \rightarrow NaCl(aq) + H_2O(l)$

Amount of HCl = amount of NaOH

$$c \times \frac{23.5}{1000} \text{ dm}^3 = 1.00 \text{ mol dm}^{-3} \times \frac{50.0}{1000} \text{ dm}^3$$

$$\therefore c = 1.00 \text{ mol dm}^{-3} \times \frac{50.0}{23.5} = \boxed{2.13 \text{ mol dm}^{-3}}$$

(b) $CH_3CO_2H(aq) + NaOH(aq) \rightarrow CH_3CO_2Na(aq) + H_2O(l)$

Amount of CH_3CO_2H = amount of NaOH

$$\therefore c \times \frac{26.5}{1000} \text{ dm}^3 = 1.00 \text{ mol dm}^{-3} \times \frac{50.0}{1000} \text{ dm}^3$$

$$\therefore c = 1.00 \text{ mol dm}^{-3} \times \frac{50.0}{26.5} = \boxed{1.89 \text{ mol dm}^{-3}}$$

(Continued on next page.)

(Continued from last page.)

4.(a) Volume of mixture when reaction is complete = $(50.0 + 23.5)cm^3$ = 73.5 cm^3

ΔT = (31.3 - 22.2)K = 9.1 K

$\Delta H = -mc_p\Delta T$ = -0.0735 kg $\times$ 4.18 kJ kg^{-1} K^{-1} $\times$ 9.1 K = -2.80 kJ

Amount of NaOH used = cV = 1.00 mol dm^{-3} $\times$ 0.0500 dm^3 = 0.0500 mol

Scaling up to 1 mol, ΔH = 2.80 kJ $\times$ $\dfrac{1}{0.0500}$ = -56.0 kJ

$\therefore$ HCl(aq) + NaOH(aq) $\rightarrow$ NaCl(aq) + H_2O(l); $\Delta H^{\ominus}$ = $\boxed{-56.0 \text{ kJ mol}^{-1}}$

(b) Volume of mixture when reaction is complete = $(50.0 + 26.5)cm^3$

$= 76.5$ cm^3

ΔT = (29.1 - 21.0)K = 8.1 K

$\Delta H = -mcp\Delta T$ = -0.0765 kg $\times$ 4.18 kJ kg^{-1} K^{-1} $\times$ 8.1 K = -2.59 kJ

Scaling up to 1 mol, ΔH = -2.59 kJ $\times$ $\dfrac{1}{0.0500}$ = 51.8 kJ

$\therefore$ CH_3CO_2H(aq) + NaOH(aq) $\rightarrow$ CH_3CO_2Na(aq) + H_2O(l);

$\Delta H^{\ominus}$ = $\boxed{-51.8 \text{ kJ mol}^{-1}}$

Experiment 14. Questions

1. The enthalpy change of neutralization for completely ionized acids and
bases is constant because the reaction is the same in every case.

$$H^+(aq) + OH^-(aq) \rightarrow H_2O(l)$$

2. Some heat is lost from the reaction mixture. The specific heat capacity of
the mixture is not precisely 4.18 kJ kg^{-1} K^{-1}.

3. The enthalpy change of neutralization for incompletely ionized acids and/or
bases is less negative because some energy is required to complete the
ionization before the reaction between hydrogen ion and hydroxide ion can
occur.

Experiment 16. Specimen results

Results Table 16

	A	B	C	D	E	F	G
Appearance	white powder	black powder	black powder	white powder	white powder	grey powder	white powder
Melting-point (approx.)/°C	> 800	> 800	> 800	100	700	> 800	> 800
Solubility in water	insoluble	insoluble	insoluble	soluble	soluble	insoluble	insoluble
Conductivity of solution	–	–	–	none	good	–	–
Conductivity of solid	none	none	good	none	none	good	none
Action of dilute HCl	none	dissolves, green solution. No H_2	none	dissolves, no evident reaction	dissolves, no evident reaction	dissolves, hydrogen evolved	none
Structure	ionic or covalent giant	ionic or covalent giant	metal or covalent giant	molecular crystal	ionic	metal	ionic or covalent giant

Notes

1. The heavy horizontal line in each column implies that you need not have done the tests below the line.

2. You were not asked to identify the actual substances. However, for your interest we list them:

   A = barium sulphate, $BaSO_4$; B = copper(II) oxide, CuO; C = graphite, C;
   D = glucose, $C_6H_{12}O_6$; E = potassium bromide, KBr; F = zinc, Zn;
   G = silica, SiO_2.

3. Your general chemical knowledge probably enabled you to guess the identity of B and C, and hence to infer the structures. The tests alone, however, are not entirely conclusive.

Experiment 16. Questions

1. The use of higher temperatures and/or more sophisticated conductivity apparatus might enable the solids with high melting points to be classified as ionic or giant covalent structures.

2. It may not be easy to get a powdered metal to conduct electricity because the particles may not make good contact with each other, especially if there is an oxide film on the surfaces.

Experiment 17. Specimen results and calculations

Results Table 17a

Room temperature	26 $^{\circ}$C
Atmospheric pressure	757 mmHg
Mass of flask filled with air	47.933 g
Mass of flask filled with CO_2	47.998 g
Mass of flask filled with water	152.8 g
Density of air under condition of experiment	0.00118 g cm^{-3}

Volume of CO_2 = volume of flask

$$= \frac{\text{mass of water}}{\text{density of water}} = \frac{152.8 \text{ g} - 47.9 \text{ g}}{1.00 \text{ g cm}^{-3}} = 104.9 \text{ cm}^3$$

Mass of empty flask = mass of flask and air - mass of air
$$= 47.933 \text{ g} - (\text{volume of air} \times \text{density of air})$$
$$= 47.933 \text{ g} - (104.9 \text{ cm}^3 \times 0.00118 \text{ g cm}^{-3})$$
$$= 47.933 \text{ g} - 0.124 \text{ g} = 47.809 \text{ g}$$

Mass of CO_2 = mass of flask and CO_2 - mass of flask
$$= 47.998 \text{ g} - 47.809 \text{ g} = 0.189 \text{ g}$$

$$pV = nRT = \frac{mRT}{M} \quad \therefore \ M = \frac{mRT}{pV}$$

If we use $R = 0.0821$ atm dm^3 K^{-1} mol^{-1},

then p must be in atm i.e. $\dfrac{757 \text{ atm}}{760 \text{ atm}} = 0.996$ atm

and V must be in dm^3 i.e. 0.105 dm^3

and T must be in K i.e. (273 + 26) K = 299 K

$$\therefore \ M = \frac{0.189 \text{ g} \times 0.0821 \text{ atm dm}^3 \text{ K}^{-1} \text{ mol}^{-1} \times 299 \text{ K}}{0.996 \text{ atm} \times 0.105 \text{ dm}^3} = \boxed{44.4 \text{ g mol}^{-1}}$$

This agrees well with the accepted value 44.1 g mol^{-1}. Your results should lie between 43 and 46 g mol^{-1}.

Experiment 17. Questions

1. If the molar mass is 44.1 g mol^{-1}, the relative molecular mass is 44.1. (This question was included simply to remind you of the difference between the two terms.)

2. Density, $\rho = \dfrac{\text{mass of sample}}{\text{volume of sample}} = \dfrac{\text{molar mass}}{\text{molar volume}} = \dfrac{M}{V_m}$

$$\rho = \frac{44.1 \text{ g mol}^{-1}}{22.4 \text{ dm}^3 \text{ mol}^{-1}} = 1.97 \text{ g dm}^{-3} \quad \text{or} \quad \boxed{1.97 \times 10^{-3} \text{ g cm}^{-3}}$$

3. Rapid removal of the delivery tube would cause turbulence and tend to reduce the pressure momentarily. This would encourage air to enter the flask.

4. The purpose of this weighing is to obtain the mass of water in the flask, which is over 100 g. An error of 0.1 g in 100 g is only 0.1%. The weighings on the more accurate balance are to obtain the mass of gas, which is only about 0.2 g. An error of 0.001 g in 0.2 g is 0.5% which is in fact _more_ serious than the error in using the less accurate balance.

Experiment 18. Specimen results

Results Table 18

Mass of hypodermic syringe and liquid before injection	13.189 g
Mass of hypodermic syringe and liquid after injection	12.920 g
Temperature of vapour	100 $^{\circ}$C
Atmospheric pressure	762 mmHg
Volume of air in syringe	5.0 cm³
Volume of air and vapour in syringe	66.0 cm³

$$pV = nRT = \frac{m}{M} RT \quad \therefore M = \frac{mRT}{pV}$$

If we use $R = 0.0821$ atm dm³ K⁻¹ mol⁻¹,

p must be in atm, i.e. $\dfrac{762 \text{ mmHg}}{760 \text{ mmHg atm}^{-1}}$ = 1.003 atm

V must be in dm³, i.e. $\dfrac{66.0 \text{ cm}^3 - 5.0 \text{ cm}^3}{1000 \text{ cm}^3 \text{ dm}^{-3}}$ = 0.0610 dm³

T must be in K, i.e. (273 + 100) K = 373 K

Then $M = \dfrac{(13.189 \text{ g} - 12.920 \text{ g}) \times 0.0821 \text{ atm dm}^3 \text{ K}^{-1} \text{ mol}^{-1} \times 373 \text{ K}}{1.003 \text{ atm} \times 0.0610 \text{ dm}^3}$

$= \boxed{135 \text{ g mol}^{-1}}$

This agrees well with the accepted value 133.4 g mol⁻¹. Although this experiment is not quite so accurate as Experiment 17, your result should lie between 130 and 140 g mol⁻¹.

Experiment 18. Questions

1. Again, the relative molecular mass is found from the molar mass by dropping the units, i.e. 135.

2. If the needle were too short, some of the liquid injected might not reach the hottest part of the syringe and therefore might not vaporise. In this case, the volume of vapour would be too small, and the final value for M would be too large. $(M = \dfrac{mRT}{pV})$

3. Volume of vapour at 762 mmHg and 373 K = 61.0 cm³

    $\therefore$ using the combined gas law,

    volume at s.t.p. = 61.0 cm³ $\times \dfrac{762 \text{ mmHg}}{760 \text{ mmHg}} \times \dfrac{273 \text{ K}}{373 \text{ K}}$ = 44.8 cm³ = 0.0448 dm³

    Scaling up to 22.4 dm³ requires a scale factor of $\dfrac{22.4}{0.0448}$

    $\therefore$ 22.4 dm³ of vapour has a mass of 0.269 g $\times \dfrac{22.4}{0.0448}$ = $\boxed{134 \text{ g}}$

4. Trichloroethane vapour is not an ideal gas, but we assume ideal behaviour both in using the ideal gas equation and also in using a fixed value for molar volume. The errors involved in making these two assumptions are fairly small but not equal. The use of the ideal gas equation is to be preferred, since we make only one assumption, namely that the gas behaves ideally at 100 °C. In using the alternative method, we have to assume ideal behaviour at 0 °C and 100 °C.

Experiment 19. Specimen results and calculations

Results Table 19

	1	2	3	4	Mean
Time for effusion of 50 cm³ hydrogen/s	10.7	10.9	10.7	10.7	10.7
Time for effusion of 50 cm³ domestic gas/s	31.3	30.9	31.0	32.1	31.3

Graham's law states:

$$\frac{\text{rate of effusion of } H_2}{\text{rate of effusion of domestic gas}} = \sqrt{\frac{\text{density of domestic gas}}{\text{density of } H_2}}$$

Since density $\propto M$ and rate of effusion $\propto \dfrac{1}{t}$ for a fixed amount, we write

$$\frac{t_{gas}}{t_{H_2}} = \sqrt{\frac{M_{gas}}{M_{H_2}}}$$

$$\therefore \quad M_{gas} = M_{H_2} \times \left(\frac{t_{gas}}{t_{H_2}}\right)^2 = 2.02 \text{ g mol}^{-1} \times \left(\frac{31.3 \text{ s}}{10.7 \text{ s}}\right)^2 = \boxed{17.3 \text{ g mol}^{-1}}$$

Your results should lie between 16.0 g mol^{-1} and 19.0 g mol^{-1}.

Experiment 19. Questions

1. The main constituent of domestic gas is methane, CH_4, which has a molar mass of 16.0 g mol^{-1}. This is consistent with results for this experiment.

2. This experiment, carefully done, nearly always gives a result greater than 16.0 g mol^{-1}. This suggests that the major impurities have molar masses greater than 16.0 g mol^{-1}. The chemical composition of domestic gas varies a little but most supplies contain well over 90% methane. The main impurities are nitrogen, and hydrocarbons such as ethane, C_2H_6, all of which have molar masses greater than 16.0 g mol^{-1}.

3. The piston should be allowed to fall from 75 cm³ to 60 cm³ before starting the stopclock to ensure that air is displaced from the tap.

4. At a higher temperature, gas molecules move more rapidly and effusion times would therefore be shorter, but in the same ratio, so that the same value for molar mass should be obtained.

5. If the hole is larger, the rate of effusion would be larger and the times shorter. Provided the hole is not so large that times are difficult to record accurately, the experiment should give the same result. Note, however, that as the hole becomes larger, Graham's law holds less well.

Experiment 20. Specimen results

Results Table 20

Change	Observation	Cause	Inference
$[Fe^{3+}]$ increased	Darker colour	Concentration of $FeSCN^{2+}$ increases	Equilibrium shifts to the right
$[SCN^-]$ increased	Darker colour	Concentration of $FeSCN^{2+}$ increases	Equilibrium shifts to the right
$[Fe^{3+}]$ decreased	Lighter colour	Concentration of $FeSCN^{2+}$ decreases	Equilibrium shifts to the left

Experiment 20. Questions

1. The position of equilibrium would shift to the left.

2. (a) The system counteracts the imposed change (an increase in $[Fe^{3+}]$) by converting some Fe^{3+} and SCN^- to $FeSCN^{2+}$.

    (b) The system counteracts the imposed change (an increase in $[SCN^-]$) by converting some SCN^- and Fe^{3+} to $FeSCN^{2+}$.

    (c) The system counteracts the imposed change (a decrease in $[Fe^{3+}]$) by converting some $FeSCN^{2+}$ to Fe^{3+} and SCN^-.

Experiment 21. Specimen results

Results Table 21a

Tube number	1A	1B	2	3	4
Mass of empty tube/g	12.01	12.24	12.57	12.18	12.73
Volume of HCl(aq) added/cm^3	5.0	5.0	5.0	5.0	5.0
Mass of tube after addition/g	17.03	17.34	17.77	17.35	17.92
Volume of ethyl ethanoate added/cm^3	——	——	5.0	4.0	2.0
Mass of tube after addition/g	——	——	22.28	20.99	19.78
Volume of water added/cm^3	——	——	——	1.0	3.0
Mass of tube after addition/g	——	——	——	21.98	22.78
Mass of ethyl ethanoate added/g	——	——	4.51	3.64	1.86
Mass of HCl(aq) added/g	——	——	5.20.	5.17	5.19
Mass of water added/g	——	——	——	0.99	3.00

Results Table 21b

Solution in flask	Equilibrium mixture				
Solution in burette	Sodium hydroxide 0.974 mol dm^{-3}				
Indicator	Phenolphthalein				
Tube number	1A	1B	2	3	4
Final burette reading	9.95	20.05	41.30	39.20	27.25
Initial burette reading	0.00	9.95	0.00	0.00	0.00
Titre/cm^3	9.95	10.10	41.30	39.20	27.35

Results Table 21c

	Tube number	2	3	4
1.	Amount of HCl/mol	0.010	0.010	0.010
2.	Total amount of acid at eqm/mol	0.041	0.039	0.027
3.	Eqm amount of ethanoic acid/mol	0.031	0.029	0.017
4.	Eqm amount of ethanol/mol	0.031	0.029	0.017
5.	Initial amount of ethyl ethanoate/mol	0.051	0.041	0.021
6.	Eqm amount of ethyl ethanoate/mol	0.020	0.012	0.004
7.	Mass of pure HCl/g	0.365	0.365	0.365
8.	Mass of water in HCl(aq)/g	4.84	4.81	4.83
9.	Initial amount of water/mol	0.269	0.322	0.435
10.	Eqm amount of water/mol	0.238	0.293	0.418
11.	Eqm constant, K_c	0.20	0.24	0.17

Experiment 22. Specimen results

Results Table 22

Solution in flask	Calcium hydroxide		? mol dm^{-3}	25.0 cm^3
Solution in burette	Hydrochloric acid		0.100 mol dm^{-3}	
Indicator	Phenolphthalein			
		Trial	1	2	3	4
Burette readings	Final		11.60	23.00	34.45	46.00
	Initial		0.00	11.60	23.00	34.50
Volume used/cm^3			11.60	11.40	11.45	11.50
Mean titre/cm^3	11.45					

Experiment 22. Calculations

1. Let A refer to H$^+$(aq) and B refer to OH$^-$(aq)

$$\frac{c_A V_A}{c_B V_B} = \frac{1}{1}$$

$$\frac{0.100 \text{ mol dm}^{-3} \times 11.45 \text{ cm}^3}{c_B \times 25.0 \text{ cm}^3} = 1$$

$$\therefore c_B = \frac{0.100 \text{ mol dm}^{-3} \times 11.45 \text{ cm}^3}{25.0 \text{ cm}^3} = 0.0458 \text{ mol dm}^{-3}$$

2. $[Ca^{2+}(aq)] = \frac{1}{2}[OH^-(aq)] = 0.0229$ mol dm^{-3}

3. Solubility of Ca(OH)$_2$ = $[Ca^{2+}(aq)]$ = 0.0229 mol dm^{-3} (at 15 $^{\circ}$C)

    Two different data books list solubilities in different ways:

    (a) 0.156 g of Ca(OH)$_2$ per 100 g of H$_2$O at 20 $^{\circ}$C

    or 1.56 g dm^3 or $\frac{1.56}{74.0}$ mol dm^{-3} = 0.0211 mol dm^{-3}

    (b) 1.53 x 10^{-3} mol per 100 g of H$_2$O $\simeq$ 0.0153 mol dm^{-3} at 25 $^{\circ}$C.

    Note that solubilities are usually quoted with reference to the volume
    of water rather than the volume of solution. This makes very little
    difference for dilute solutions such as saturated calcium hydroxide.
    Also, note that calcium hydroxide unlike most salts, becomes less
    soluble at higher temperatures.

4. (a) K_s = $[Ca^{2+}(aq)][OH^-(aq)]^2$

    = 0.0229 mol dm^{-3} $\times$ (0.0458 mol dm^{-3})^2 = 4.80 $\times$ 10^{-5} mol^3 dm^{-9}

    (b) K_s = 0.0211 mol dm^{-3} $\times$ (0.0422 mol dm^{-3})^2 = 3.76 $\times$ 10^{-5} mol^3 dm^{-9}

    or K_s = 0.0153 mol dm^{-3} $\times$ (0.0306 mol dm^{-3})^2 = 1.43 $\times$ 10^{-5} mol^3 dm^{-9}

    (Another data book lists the solubility product at 25 $^{\circ}$C as
    5.5 x 10^{-6} mol^3 dm^{-9} Clearly there is a considerable degree of
    uncertainty.)

5. Absorption of CO$_2$ from the air reduces $[OH^-(aq)]$.

Experiment 23. Questions

1. A colourless precipitate appeared in each tube.

2. The system counteracts an increase in the concentration of either of
    the ions by shifting the equilibrium position to the left (i.e.
    producing solid NaCl)

$$NaCl(s) \rightleftharpoons Na^+(aq) + Cl^-(aq)$$

3. Each ion has concentration = 5.42 mol dm^{-3}.

4. Some of the sodium ions in solution interacted with the excess
    chloride ions to form a precipitate of sodium chloride.

5. Some of the chloride ions in solution interacted with the excess
    sodium ions to form a precipitate of sodium chloride.

6. At high concentrations, ions of opposite charge interact with each
    other, tending to 'pair up'. This makes the effective concentration
    of 'free' ions very much smaller than the concentrations obtained from
    the amount dissolved. The equilibrium law can only be applied using
    concentrations if these are the same as the effective concentrations
    of 'free' ions, which is the case in very dilute solutions only.
    (Effective concentrations are known as 'activities' and, if values can
    be calculated, they can be used in the equilibrium law, but this is
    beyond A-level.)

Experiment 24. Specimen results and calculations

Results Table 24a

Solution in flask	NH_3 in CH_3CCl_3			? mol dm^{-3}	10 cm^3	
Solution in burette	HCl(aq)			0.010 mol dm^{-3}		
Indicator	methyl orange					
		Trial	1	2	3	4
Burette readings	Final		4.60	4.60	9.05	
	Initial		0.00	9.05	13.60	
Volume used/cm^3			4.60	4.45	4.55	
Mean titre/cm^3	4.50					

Results Table 24b

Solution in flask	NH_3(aq)			? mol dm^{-3}	10 cm^3	
Solution in burette	HCl(aq)			0.50 mol dm^{-3}		
Indicator	methyl orange					
		Trial	1	2	3	4
Burette readings	Final		25.90	25.65	25.85	
	Initial		00.00	00.00	00.00	
Volume used/cm^3			25.90	25.65	25.85	
Mean titre/cm^3	25.80					

Experiment 24. Calculations

1. Substituting in the expression

$$\frac{c_A V_A}{c_B V_B} = \frac{a}{b}$$

where A refers to HCl and B to NH_3

gives $\dfrac{0.010 \text{ mol dm}^{-3} \times 4.50 \text{ cm}^3}{c_B \times 10.0 \text{ cm}^3} = \dfrac{1}{1}$

$\therefore \quad c_B = 0.010 \text{ mol dm}^{-3} \times \dfrac{4.50}{10.0} = \boxed{4.5 \times 10^{-3} \text{ mol dm}^{-3}}$

2. By the same method as in 1.

$c_B = 0.50 \text{ mol dm}^{-3} \times \dfrac{25.8}{10.0} = \boxed{1.3 \text{ mol dm}^{-3}}$

3. $K_D = \dfrac{[NH_3(aq)]}{[NH_3(tce)]} = \dfrac{1.3 \text{ mol dm}^{-3}}{4.5 \times 10^{-3} \text{ mol dm}^{-3}} = \boxed{2.9 \times 10^2}$

Experiment 25. Specimen results

Results Table 25

Concentration of acid mol/dm^{-3}	Observed pH of solutions of ethanoic acid	Calculated pH of solutions of hydrochloric acid
0.00010	4.2	4.0
0.0010	3.5	3.0
0.010	3.0	2.0
0.10	2.7	1.0

Experiment 25. Questions

1. (a) In each case, the hydrochloric acid solution has a higher concentration of hydrogen ions than the ethanoic acid solution of the same concentration.

   (b) This shows that the ethanoic acid solutions are incompletely dissociated.

2. (a) As the ethanoic acid is diluted, its pH becomes closer and closer to the pH of the strong acid at the same concentration. This shows that dilution increases the extent of dissociation of ethanoic acid. The same is true of other weak acids.

   The trend is continued at greater dilution, but precise measurement becomes much more difficult due to contamination of the sample with dissolved CO_2.

   However, calculations show that even at a concentration of 10^{-6} mol dm^{-3}, the ethanoic acid would not be quite fully dissociated; it would have a pH of 6.02 as compared with pH of 6.00 for a strong acid.

2. (b) The full equation for the dissociation of ethanoic acid in water is:

$$CH_3CO_2H(aq) + H_2O(l) \rightleftharpoons CH_3CO_2^-(aq) + H_3O^+(aq)$$

   The addition of water decreases the concentrations of ethanoic acid, ethanoate ions and hydrogen ions. Because two of these are on the right of the equation and only one on the left, a shift to the right goes further towards restoring the original concentrations than a shift to the left. (The decrease in concentration reduces the rate of reaction R → L to a greater extent than L → R.) Thus, the acid becomes increasingly dissociated on dilution.

   Note that the alternative simpler argument, that the added water is partially used up in a shift to the right, is invalid because the concentration of water remains effectively constant.

Experiment 26. Specimen results

Results Table 26

Acid	Concentration /mol dm^{-3}	pH
Boric acid	0.010	5.6
Benzoic acid	0.010	2.8
Ethanoic acid	0.010	3.0
Dihydrogenphosphate(V) ion	0.010	4.6

Experiment 26. Questions

Order of acid strength, strongest acid first:

benzoic acid > ethanoic acid > dihydrogenphosphate(V) ion > boric acid

Experiment 27. Specimen results

Results Table 27

Volume added	pH on addition of 0.1 M NaOH to		pH on addition of 0.1 M HCl to	
	buffer	pure water	buffer	pure water
0	7.0	6.9	7.0	6.8
1 drop	7.0	9.8	7.0	4.1
1.0 cm^3	7.1	10.5	6.9	2.7
5.0 cm^3	10.0	11.6	6.0	2.1
pH of pure water with minimum air exposure	6.9			
pH of pure water after 10 mins air exposure	5.4			

Experiment 27. Questions

1. You will probably record a change of about 4 pH units. Theoretically, the change should be from pH 7.0 to pH 2.4 so, if everything is right in your experiment, you might answer 5 pH units.

2. Assuming that the pH of pure water after the addition of acid was 3.0, to the nearest unit,

$$[H^+(aq)] = 1.0 \times 10^{-3} \text{ mol dm}^{-3}$$

$[H^+(aq)]$ in pure water (pH 7.0) $= 1.0 \times 10^{-7}$ mol dm^{-3}

$$\therefore \frac{[H^+(aq)] \text{ after addition of acid}}{[H^+(aq)] \text{ in pure water}} = \frac{1.0 \times 10^{-3} \text{ mol dm}^{-3}}{1.0 \times 10^{-7} \text{ mol dm}^{-3}} = 10^4$$

In other words, a decrease of 4 pH units means there are 10^4 (or 10,000) times as many hydrogen ions in the same volume.

3. Almost all the hydrogen ions must have reacted with the buffer solution. You will find out what this reaction is later.

4. Carbon dioxide is absorbed from the air and reacts with water to form hydrogen ions.

$$CO_2(aq) + H_2O(l) \rightleftharpoons H^+(aq) + HCO_3^-(aq)$$

The distilled water you normally use in the laboratory is pure enough for most purposes, but nevertheless it rapidly becomes somewhat acidic.

Experiment 28. Specimen results and questions

1. From the substances available, the acid with a pK_a value nearest to the
    required pH of 5.2 is ethanoic acid, pK_a = 4.8.

$$pH = pK_a - \log \frac{[CH_3CO_2H(aq)]}{[CH_3CO_2^-(aq)]}$$

$$\therefore \frac{[CH_3CO_2H(aq)]}{[CH_3CO_2^-(aq)]} = \text{antilog } (pK_a - pH) = \text{antilog } (4.8 - 5.2)$$

$$= \text{antilog } (-0.4) = 0.40$$

If x cm³ of 1.0 M CH_3CO_2H is used to make 100 cm³ of buffer,
(100 - x)cm³ of 1.0 M CH_3CO_2Na is required.

$$\frac{x}{100 - x} = 0.4 \quad \text{or} \quad x = 40 - 0.4\,x$$

$$x = \frac{40}{1.4} = 29$$

$\therefore$ | use 29 cm³ of 1.0 M CH_3CO_2H and 71 cm³ of 1.0 M CH_3CO_2Na |

2. From the substances available, the acid with a pK_a value nearest to the
    required pH of 8.8 is the ammonium ion, pK_a = 9.3.

$$\frac{[NH_4^+(aq)]}{[NH_3(aq)]} = \text{antilog } (pK_a - pH) = \text{antilog } (9.3 - 8.8)$$

$$= \text{antilog } (0.5) = 3.2$$

If x cm³ of 1.0 M NH_4Cl is used to make 100 cm³ of buffer, (100 - x) cm³
of 1.0 M NH_3 is required.

$$\frac{x}{100 - x} = 3.2 \quad \text{or} \quad x = 320 - 3.2\,x$$

$$x = \frac{320}{4.2} = 76$$

$\therefore$ | use 76 cm³ of 1.0 M NH_4Cl and 24 cm³ of 1.0 M NH_3. |

3. Measured values should be within 0.2 units of the required values.
    Differences may be due to inaccurate pH meters but, in any case, the
    solutions used are too concentrated for the equilibrium law to be applied
    precisely using concentrations. pH meters actually measure -log(activity
    of hydrogen ions), where 'activity' is the technical term for 'effective
    concentration'. At high concentrations, ions interact with each other so
    that their effective concentration is reduced - as a result, pH meters
    tend to record higher values than you might expect for concentrated
    solutions.

4. The equation used in the calculation suggests that the pH depends only
    on the ratio of concentrations of acid and conjugate base, so that
    dilution should have no effect. In practice, the pH of these buffers
    falls a little with dilution - between 0.1 and 0.2 units for a 10:1
    dilution. Again, this is due to the fact that measured pH depends on
    activity, which changes with concentration, and also to the breakdown of
    the assumptions made in the calculation, e.g.

$$[CH_3COOH(aq)] \simeq \text{molarity in dilute solutions.}$$

5. The capacity of the buffers falls dramatically with dilution. The
    concentrated buffers will absorb 10 cm³ of 0.2 M HCl or 0.2 M NaOH with
    a pH change of only 0.1 or 0.2 units. A similar change for buffers
    diluted by 10:1 is brought about by only 1 cm³ of acid or alkali; while
    for buffers diluted by 100:1 a single drop of acid or alkali is enough.

6. Ammonia is so volatile that the composition of the buffer will not
    remain constant.

Experiment 29. Specimen results

Results Table 29a

Indicator	Initial colour	Colour change starts		Colour change ends		Final colour
		Volume	pH	Volume	pH	
Litmus	red	8 cm³	4.9	24 cm³	8.5	blue
Methyl orange	red	1 cm³	3.0	11 cm³	5.6	yellow
Phenolphthalein	colourless	22 cm³	8.0	30 cm³	9.9	red
Bromophenol blue	yellow	1 cm³	3.0	9 cm³	5.1	blue
Methyl red	red	8 cm³	4.9	15 cm³	6.4	yellow

Table 29b

indicator — pH range

- alizarin yellow: yellow / red
- phenolphthalein: colourless / red
- phenol red: yellow / red
- litmus: red / blue
- methyl red: red / yellow
- bromophenol blue: yellow / blue
- thymol blue: red / yellow

pH: 0 1 2 3 4 5 6 7 8 9 10 11 12 13 14

Experiment 29. Questions

The pH was measured in a buffer solution to ensure that the pH did not change
too rapidly on the addition of alkali. It would be almost impossible to
change the pH of the solution in a controlled way without using a buffer.

Experiment 30. Specimen results and calculations

Method 1

Best colour match with tubes corresponding to $\dfrac{[HIn(aq)]}{[In^-(aq)]} = \dfrac{7}{3}$

$$pH = pK_{In} - \log \dfrac{[HIn(aq)]}{[In^-(aq)]}$$

$$\therefore pK_{In} = pH + \log \dfrac{[HIn(aq)]}{[In^-(aq)]} = 3.7 + \log \dfrac{7}{3} = 3.7 + 0.37 = 4.1$$

$$K_{In} = \text{antilog}\,(-pK_{In}) = \text{antilog}\,(-4.1) = \boxed{7.9 \times 10^{-5} \text{ mol dm}^{-3}}$$

Method 2

Best colour match when $\dfrac{[HIn(aq)]}{[In^-(aq)]} = \dfrac{3.0 \text{ cm}}{1.6 \text{ cm}} = 1.9$

$$pH = pK_{In} - \log \dfrac{[HIn(aq)]}{[In^-(aq)]}$$

$$\therefore pK_{In} = pH + \log \dfrac{[HIn(aq)]}{[In^-(aq)]} = 3.7 + \log 1.9 = 3.7 + 0.28 = 4.0$$

$$K_{In} = \text{antilog}\,(-pK_{In}) = \text{antilog}\,(-4.0) = \boxed{1.0 \times 10^{-4} \text{ mol dm}^{-3}}$$

Experiment 30. Questions

1. The wedge method gives a continuous range of colours, which makes
    comparison easier, and avoids the difficulty of looking through curved
    surfaces. However, the apparatus is more expensive and requires larger
    volumes of solutions.

2. The total concentrations of indicator, $[HIn(aq)]$ and $[In^-(aq)]$ had to
    be the same in the solutions being compared, otherwise a change in
    density might be confused with a change in colour.

3. A buffer of pH 8 could not be used since pH 8 is outside the pH range
    of the indicator. Colour comparisons can only be made when the ratio
    $[HIn(aq)]/[In^-(aq)]$ is between 10 and 0.1 (approximately).

Experiment 31. Specimen results and calculations

Method 1
Best colour match with tubes corresponding to $\dfrac{[HIn(aq)]}{[In^-(aq)]} = \dfrac{4}{6}$

$$\therefore pH = pK_{In} - \log \dfrac{[HIn(aq)]}{[In^-(aq)]} = 4.0 - \log \dfrac{4}{6} = 4.2$$

$$pH = pK_a - \log \dfrac{[C_6H_5CO_2H(aq)]}{[C_6H_5CO_2^-(aq)]}$$

$$4.2 = pK_a - 0 \qquad (\text{since } [C_6H_5CO_2H(aq)] = [C_6H_5CO_2^-(aq)])$$

$$\therefore pK_a = 4.2 \text{ and } K_a = \text{antilog}\,(-4.2) = \boxed{6.3 \times 10^{-5} \text{ mol dm}^{-3}}$$

Method 2
Best colour match when $\dfrac{[HIn(aq)]}{[In^-(aq)]}$ corresponds to $\dfrac{2.0 \text{ cm}}{3.0 \text{ cm}}$

$$\therefore pH = pK_{In} - \log \dfrac{[HIn(aq)]}{[In^-(aq)]} = 4.0 - \log \dfrac{2}{3} = 4.2$$

$$pH = pK_a - \log \dfrac{[C_6H_5CO_2H(aq)]}{[C_6H_5CO_2^-(aq)]}$$

$$4.2 = pK_a - 0 \qquad (\text{since } [C_6H_5CO_2H(aq)] = [C_6H_5CO_2^-(aq)])$$

$$\therefore pK_a = 4.2 \text{ and } K_a = \text{antilog}\,(-4.2) = \boxed{6.3 \times 10^{-5} \text{ mol dm}^{-3}}$$

Experiment 32. Specimen results

Results Table 32

0.100 M NH$_3$ added to 25.0 cm^3 of 0.100 M CH$_3$CO$_2$H

Volume/cm^3	0.0	5.0	10.0	15.0	20.0	24.0	24.2	24.4	24.6	24.8	24.9
pH	3.0	4.2	4.6	4.9	5.3	6.2	6.3	6.4	6.6	6.7	6.8
Volume/cm^3	24.95	25.0	25.05	25.1	25.2	25.4	25.6	25.8	26.0	30.0	35.0
pH	6.9	7.0	7.1	7.2	7.3	7.4	7.6	7.7	7.8	8.6	8.9

0.100 M NaOH added to 25.0 cm^3 of 0.100 M CH$_3$CO$_2$H

Volume/cm^3	0.0	5.0	10.0	15.0	20.0	24.0	24.2	24.4	24.6	24.8	24.9
pH	3.0	4.2	4.6	4.9	5.3	6.2	6.4	6.6	6.9	7.2	7.6
Volume/cm^3	24.95	25.0	25.05	25.1	25.2	25.4	25.6	25.8	26.0	30.0	35.0
pH	8.1	9.0	9.5	10.0	10.6	10.9	11.2	11.3	11.4	11.9	12.2

0.100 M NH$_3$ added to 25.0 cm^3 of 0.100 M HCl

Volume/cm^3	0.0	5.0	10.0	15.0	20.0	24.0	24.2	24.4	24.6	24.8	24.9
pH	1.0	1.2	1.4	1.6	2.0	2.7	2.9	3.1	3.3	3.8	4.6
Volume/cm^3	24.95	25.0	25.05	25.1	25.2	25.4	25.6	25.8	26.0	30.0	35.0
pH	5.3	6.0	6.7	7.1	7.3	7.5	7.6	7.7	7.8	8.6	8.9

0.100 M NaOH added to 25.0 cm^3 of 0.100 M HCl

Volume/cm^3	0.0	5.0	10.0	15.0	20.0	24.0	24.2	24.4	24.6	24.8	24.9
pH	1.0	1.2	1.4	1.6	2.0	2.7	3.0	3.5	4.0	4.6	5.0
Volume/cm^3	24.95	25.0	25.05	25.1	25.2	25.4	25.6	25.8	26.0	30.0	35.0
pH	5.2	7.0	8.9	9.8	10.6	10.9	11.2	11.3	11.4	11.9	12.2

Experiment 32. Questions

(All the numerical answers in the table below assume that the acids and alkalis were precisely 0.100 M)

	0.10 M CH$_3$CO$_2$H 0.10 M NH$_3$	0.10 M CH$_3$CO$_2$H 0.10 M NaOH	0.10 M HCl 0.10 M NH$_3$	0.10 M HCl 0.10 M NaOH
Q1	NONE	pH 7.4 - 10.2	pH 3.6 - 6.8	pH 3.6 - 10.2
Q2	25.0 cm^3	25.0 cm^3	25.0 cm^3	25.0 cm^3
Q3	0, 0.1, 0	0, 2.5, 0	0, 2.5, 0	0, 6.0, 0
Q4	NONE	PHENOLPHTHALEIN thymol red thymol blue thymolphthalein bromothymol blue phenol red	METHYL ORANGE bromophenol blue congo red bromocresol green methyl red bromocresol purple	all mentioned in previous columns, plus litmus

Experiment 33. Specimen results

Results Table 33b

Volume/cm³		% composition	Boiling-point/°C		
A	B	A (by volume)	System 1	System 2	System 3
10	0	100	61.2	77	97.4
10	2	83.3	64.0	70	94.8
10	4	71.4	65.2	67.2	93.0
10	6	62.5	64.8	66	91.6
10	8	55.6	64.6	65.2	90.8
10	10	50	64.4	64.8	90.0
0	10	0	58.0	81	82.6
2	10	16.7	60.0	65.4	85.0
4	10	28.6	62.0	64.6	86.6
6	10	37.5	64.0	64.8	87.8
8	10	44.4	64.2	65.0	88.8

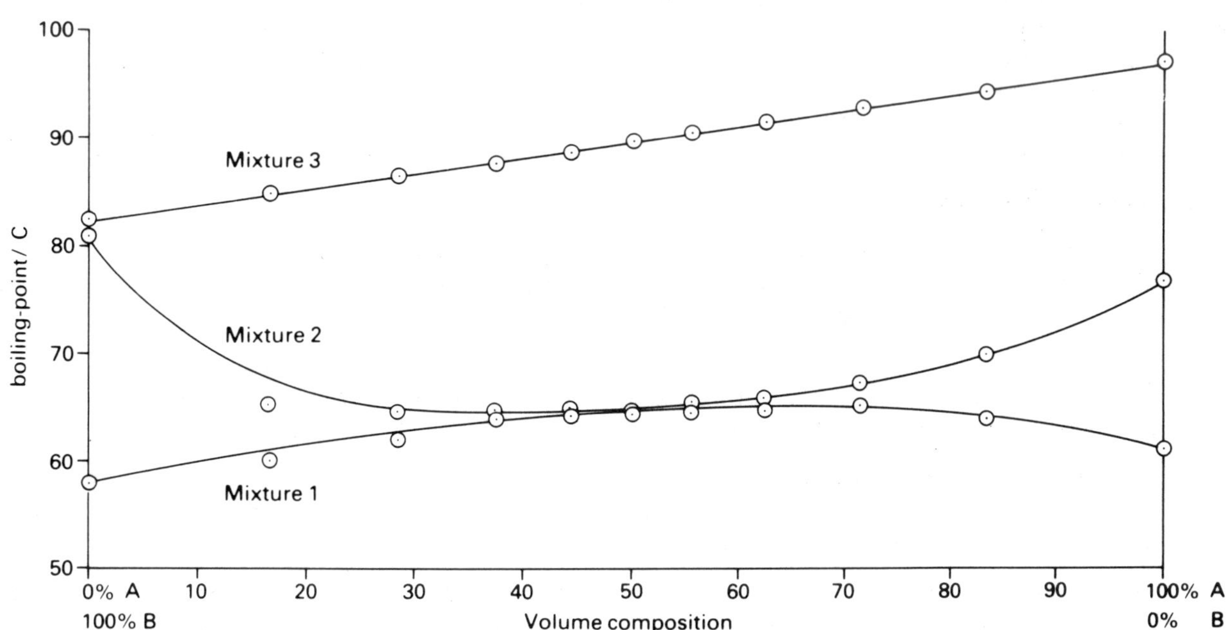

Experiment 33. Questions

1.

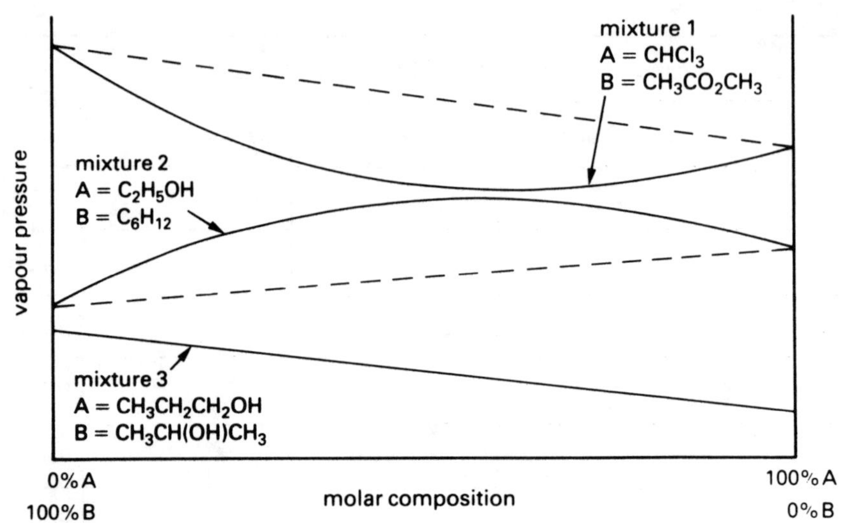

2. Mixture 1 (trichloromethane and methyl ethanoate) shows a negative
    deviation from ideality.

    Mixture 2 (ethanol and cyclohexane) shows a positive deviation.

    Mixture 3 (propan-1-ol and propan-2-ol) is virtually ideal.

Experiment 34. Specimen results

Results Table 34

Mixture	$T_A/°C$	$T_B/°C$	$T_1/°C$	$T_2/°C$	$\Delta T/°C$
1. Trichloromethane/methyl ethanoate	23.0	23.0	23.0	32.0	+9.0
2. Ethanol/cyclohexane	24.0	24.0	24.0	20.5	-3.5
3. Propan-1-ol/propan-2-ol	24.0	24.5	24.5	24.5	0

Experiment 34. Questions

1. Mixture 1: ΔT is positive, $\therefore$ $\Delta H^{\ominus}_{mix}$ is negative.
    Mixture 2: ΔT is negative, $\therefore$ $\Delta H^{\ominus}_{mix}$ is positive.
    Mixture 3: ΔT is zero, $\therefore$ $\Delta H^{\ominus}_{mix}$ is zero.

2. Mixture 3, propan-1-ol and propan-2-ol, is virtually ideal. There is no
    energy change on mixing, which shows that the A—B bonds formed are of the
    same strength as the A—A and B—B bonds which they replace.

3. The temperature increases when trichloromethane and methyl ethanoate are
    mixed because more energy is released in making A—B bonds than is required
    to break the A—A and B—B bonds which they replace.

4. Since molecules of trichloromethane and methyl ethanoate are more strongly
    bound together in the mixture than molecules in the pure liquids, their
    tendency to escape into the vapour phase is reduced. This causes a reduc-
    tion in vapour pressure compared with an ideal mixture. The effect is
    great enough to cause a minimum in the vapour pressure/composition curve.

5. The positive value of $\Delta H^{\ominus}_{mix}$ for the ethanol-cyclohexane system shows that
    A—B bonds are weaker than the A—A and B—B bonds. This gives a higher
    vapour pressure than ideal and the effect is great enough to give a
    maximum in the vapour pressure/composition curve.

Experiment 35. Specimen results and calculations

Mixing 30 cm^3 of methyl ethanoate with 3.0 cm^3 of trichloromethane gave a temperature rise of 1.5 ^{o}C.

Mass of methyl ethanoate = 30 cm^3 x 0.93 g cm^{-3} = 27.9 g

Amount of methyl ethanoate = 27.9 g/74.1 g mol^{-1} = 0.377 mol

Heat capacity of methyl ethanoate = 27.9 g x 1.97 J g^{-1} K^{-1} = 55.0 J K^{-1}

Mass of trichloromethane = 3.0 cm^3 x 1.49 g cm^{-3} = 4.47 g

Amount of trichloromethane = 4.47 g/119.4 g mol^{-1} = 0.0374 mol

Heat capacity of trichloromethane = 4.47 g x 0.96 J g^{-1} K^{-1} = 4.3 J K^{-1}

Mass of boiling-tube = 28.7 g

Heat capacity of boiling-tube = 28.7 g x 0.67 J g^{-1} K^{-1} = 19.2 J K^{-1}

$\therefore$ heat released when 0.0374 mol of CHCl$_3$ forms hydrogen bonds is given by:

   1.5 K x (55.0 + 4.3 + 19.2)J K^{-1} = 118 J = 0.118 kJ

$\therefore$ hydrogen-bond energy = $\dfrac{0.118 \text{ kJ}}{0.0374 \text{ mol}}$ = $\boxed{3.2 \text{ kJ mol}^{-1}}$

Experiment 36. Specimen results

Results Table 36

Liquid	Formula	Time/s
Propan-1-ol	CH$_3$CH$_2$CH$_2$OH	13
Propane-1,2-diol	CH$_3$CH(OH)CH$_2$OH	31
Propane-1,2,3-triol	CH$_2$(OH)CH(OH)CH$_2$OH	800
Propane-1,2,3-triyl triethanoate	CH$_2$(OCOCH$_3$)CH(OCOCH$_3$)CH$_2$OCOCH$_3$	17

Experiment 36. Questions

1. Each alcohol molecule forms hydrogen bonds with its neighbours through hydroxyl groups ($-$O$-$H$\cdots$O$<$). As the number of hydroxyl groups increases, so does the extent of hydrogen bonding in the liquid; this results in a great increase in viscosity.

2. Propane-1,2,3-triyl triethanoate has the stronger intermolecular van der Waals forces because of the greater number of electrons, but it cannot form hydrogen bonds because there are no hydrogen atoms with a sufficient positive charge. Propane-1,2,3-triol has extensive hydrogen bonding and this outweighs the effect of the weaker van der Waals forces.

Experiment 37. Specimen results

Results Table 37

Compound	Formula	Structural formula	Deflection
Cyclohexane	C_6H_{12}		0
Cyclohexene	C_6H_{10}		0 - 1
Hexane	C_6H_{14}	H—C—C—C—C—C—C—H (with H above and below each C)	0
Hexan-1-ol	$C_6H_{13}OH$	H—C—C—C—C—C—C—OH (with H above and below each C)	3
Methyl ethanoate	$CH_3CO_2CH_3$	H—C—C(=O)—O—CH_3	3
Propanone	CH_3COCH_3	H—C—C(=O)—CH_3	2
Tetrachloromethane	CCl_4	Cl—C—Cl (with Cl above and below)	0
Trichloromethane	$CHCl_3$	Cl—C—H (with Cl above and below)	1
Water	H_2O	H—O—H	2

Experiment 37. Questions

1. The charged rod attracts that part of each polar molecule which has an opposite charge and repels that part with the same charge. This causes the molecules to align themselves along the lines of force and to move towards the charged rod.

(Continued on next page.)

(Continued from last page.)

2. The molecules align themselves in the opposite sense but the observed
    movement is still in the same direction as before.

3. The approximate order of decreasing polarity is:

    water, trichloromethane, hexan-1-ol, methyl ethanoate, propanone,
    cyclohexene, followed by the three non-polar liquids, cyclohexane,
    hexane and tetrachloromethane.

    This order reflects both the relative polarity of bonds in the molecule
    and the degree of symmetry,

    (a) Both trichloromethane and tetrachloromethane
         contain highly polar C—Cl bonds, but in the latter
         case they are arranged symmetrically so that their
         effects cancel, making the molecule as a whole non-
         polar. Trichloromethane is strongly polar.

    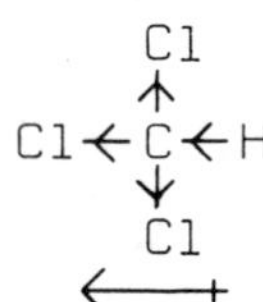

    (b) The cyclohexane molecule is symmetrical and is
         therefore non-polar. Cyclohexene is non-symmetrical;
         the net effect of the slightly polar C—H bonds makes
         the molecule slightly polar.

    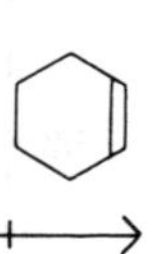

Experiment 38. Specimen results and calculations

Results Table 38a

	Experiment 1	Experiment 2
Mass of boiling-tube + B/g	31.74	29.34
Mass of boiling-tube/g	30.96	27.80
Mass of B/g	0.78	1.54
Mass of boiling-tube + A + B/g	49.17	47.18
Mass of A/g	17.43	17.84

Results Table 38b Experiment 1

Time/min	0	$\frac{1}{2}$	1	$1\frac{1}{2}$	2	$2\frac{1}{2}$	3	$3\frac{1}{2}$	4
Temp./°C	57.25	55.5	54.0	52.75	51.5	50.5	50.25	50.25	50.25
Time/min		$4\frac{1}{2}$	5	$5\frac{1}{2}$	6	$6\frac{1}{2}$	7	$7\frac{1}{2}$	8
Temp./°C		50.25	50.25	50.25	50.25	50.25	50.0	50.0	50.0

(Continued on next page.)

(Continued from last page)

Results Table 38c Experiment 2

Time/min	0	$\frac{1}{2}$	1	$1\frac{1}{2}$	2	$2\frac{1}{2}$	3	$3\frac{1}{2}$	4
Temp./°C	57.25	56.0	54.75	53.5	52.25	51.25	50.25	49.5	48.5
Time/min		$4\frac{1}{2}$	5	$5\frac{1}{2}$	6	$6\frac{1}{2}$	7	$7\frac{1}{2}$	8
Temp./°C		47.75	48.25	48.25	48.25	48.25	48.0	48.0	48.0

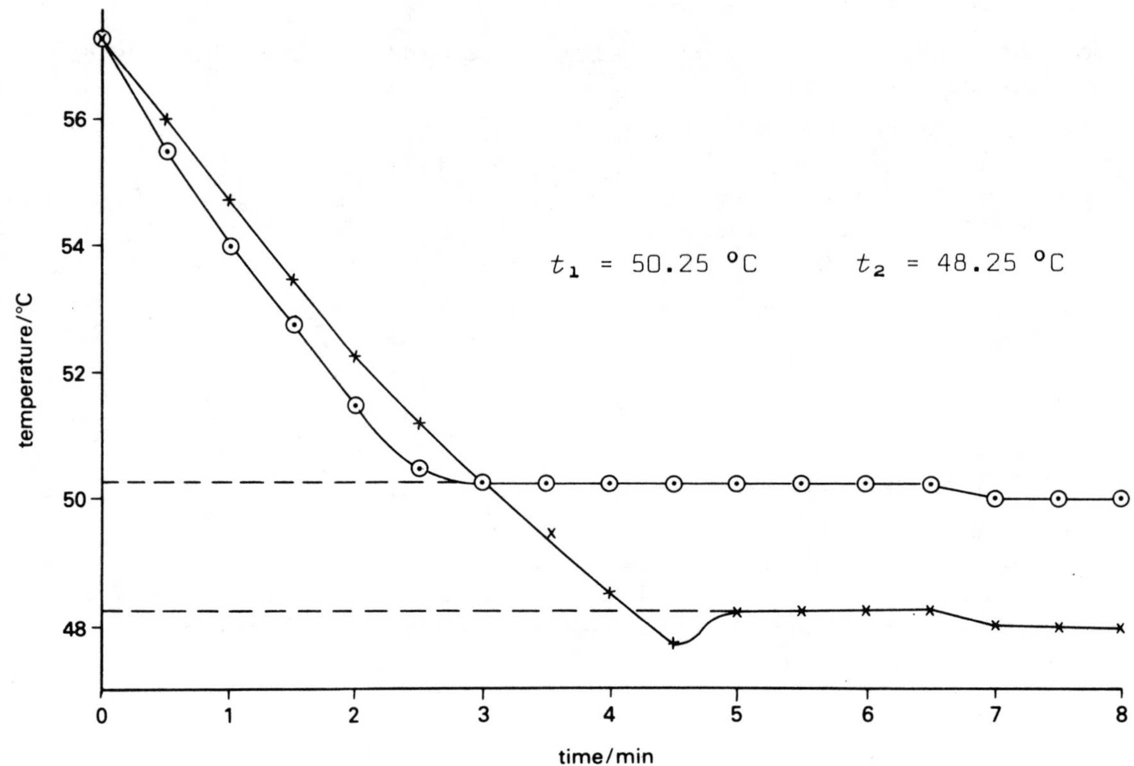

Calculation

Experiment 1. $M_r = \dfrac{7.10 \times 0.78 \times 10^3}{(53.0 - 50.25) \times 17.43} = \boxed{116}$

Experiment 2. $M_r = \dfrac{7.10 \times 1.54 \times 10^3}{(53.0 - 58.25) \times 17.84} = \boxed{129}$

Mean value of $M_r = \dfrac{116 + 129}{2} = 122.5$

Rearranging the equation for M gives:

$$53 - t(°C) = \frac{7.10 \times \text{mass of } B(g) \times 10^3}{M \times \text{mass of } A(g)}$$

$$= \frac{7.10 \times (0.78 + 1.54) \times 10^3}{122.5 \times 17.43} = 7.7$$

$$\therefore \quad t = (53.0 - 7.7) °C = \boxed{45.3 °C}$$

An approximate answer is given by simply adding the freezing-point
depressions from each experiment (2.75 + 4.75 = 7.5 °C) and
subtracting this from 53 °C. (This answer is approximate because the
masses of A used in Experiments 1 and 2 were not quite equal.)

Experiment 39. Specimen results

Results Table 39b

| Salt | Mass/g | | | Temperature/$^{\circ}$C | | | ΔH_{soln}/kJ mol^{-1} = total mass (kg) $\times$ ($-\Delta T$) $\times$ 10 $\times$ 4.18 kJ kg^{-1} K^{-1} |
	Empty bottle	Bottle + salt	Salt	Initial	Final	Change	
LiCl	5.62	9.86	4.24	22.0	34.5	+12.5	-28.3
NaCl	5.73	11.57	5.84	22.0	20.5	-1.5	+3.50
KCl	5.68	13.14	7.46	22.5	15.0	-7.5	+18.0
CaCl$_2$	5.61	16.71	11.10	23.0	51.0	+28.0	-71.5
FeCl$_3$	5.68	21.88	16.20	23.0	60.0	+37.0	-102

Results Table 39c

| Salt | Mass of salt/g | Volume of salt, V_1/cm^3 | Burette reading | | % volume change, $\dfrac{V_3 - V_2 - V_1}{50 + V_1} \times 100$ |
			Initial, V_2/cm^3	Final, V_3/cm^3	
LiCl	4.24	2.02	46.20	47.50	-1.4
NaCl	5.84	2.65	46.20	48.55	-0.6
KCl	7.46	3.73	46.20	49.30	-1.2
CaCl$_2$	11.10	4.44	46.20	48.70	-3.6
FeCl$_3$	16.20	5.79	46.20	50.45	-2.8

Experiment 39. Questions

1. ΔH_{soln} is expressed per mole of solute. 0.10 mol of solute was used in the experiment; therefore, the value of the heat energy released must be multiplied by 10.

2. (a) ΔH_{soln} becomes less negative (i.e. the process is less exothermic) going down the group.

   (b) ΔH_{soln} becomes less negative as the ionic radius increases.

3. (a) ΔH_{soln} becomes markedly more negative from KCl to $FeCl_3$.

   (b) The increased exothermicity appears to be greater than would be expected on grounds of decreased size of cation and increased number of chloride ions. Therefore, it seems that an increase in the charge of the cation makes ΔH_{soln} more negative.

4. The fact that ΔH_{soln} is negative for $CaCl_2$ implies that the energy released when ions interact with water molecules must be greater than the sum of the energies required to separate water molecules from each other and to separate ions in the solid.

5. (a) The reduction in volume must be due to a closer packing of water molecules around the ions than occurs in pure water.

   (b) Broadly speaking, the volume change is greater for more negative values of ΔH_{soln}. This is probably due to the fact that small, highly-charged ions attract water molecules very strongly. A larger number of molecules is packed tightly around the ion, giving a greater volume reduction and a greater release of energy.

Experiment 40. Specimen results and calculations

Results Table 40a

Time, t/min	0	1	2	3	4	5	6
Volume of N_2, V_t/cm³	0	14	28	41	54	65	76
$(V_\infty - V_t)$/cm³	219	205	191	178	165	154	143
Time, t/min	7	8	9	10	11	12	13
Volume of N_2, V_t/cm³	87	96	104	112	120	127	133
$(V_\infty - V_t)$/cm³	132	123	115	107	99	92	86
Time, t/min	14	15	16	17	18	19	20
Volume of N_2, V_t/cm³	139	145	150	155	160	164	168
$(V_\infty - V_t)$/cm³	80	74	69	64	59	55	51
Time, t/min	21	22	23	24	25	∞	
Volume of N_2, V_t/cm³	172	175	177	181	184	219	
$(V_\infty - V_t)$/cm³	47	44	42	38	35	0	

Temperature of thermostat: 47 °C

(Continued on next page.)

(Continued from last page.)

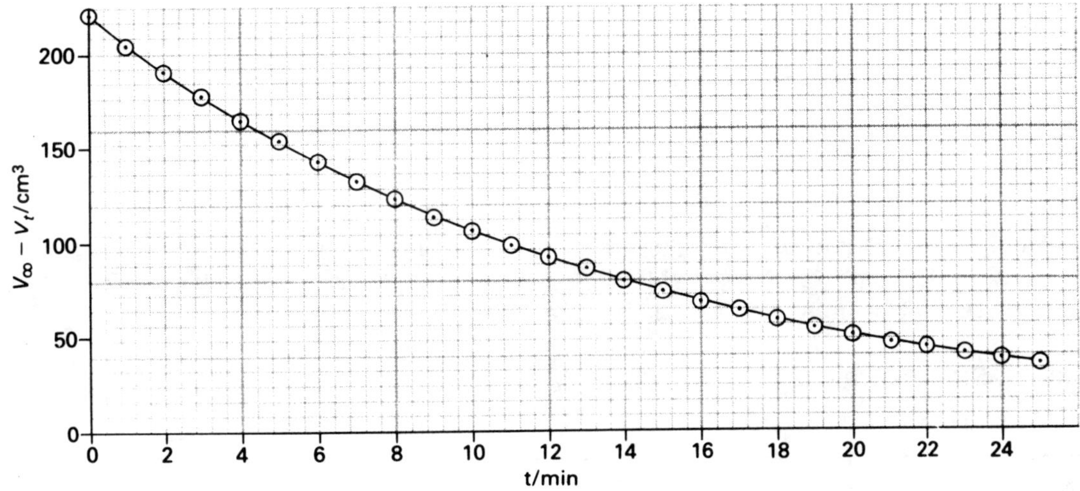

Results Table 40b

Time /min	Slope /cm³ min⁻¹	Rate /cm³ min⁻¹	$\left(V_\infty - V_t\right)$ /cm³
0	$\dfrac{219}{13.3}$	16.5	219
4	$\dfrac{214}{17.7}$	12.1	165
7	$\dfrac{202}{20.1}$	10.0	132
14	$\dfrac{150}{24.1}$	6.22	80
21	$\dfrac{88}{22.9}$	3.84	47

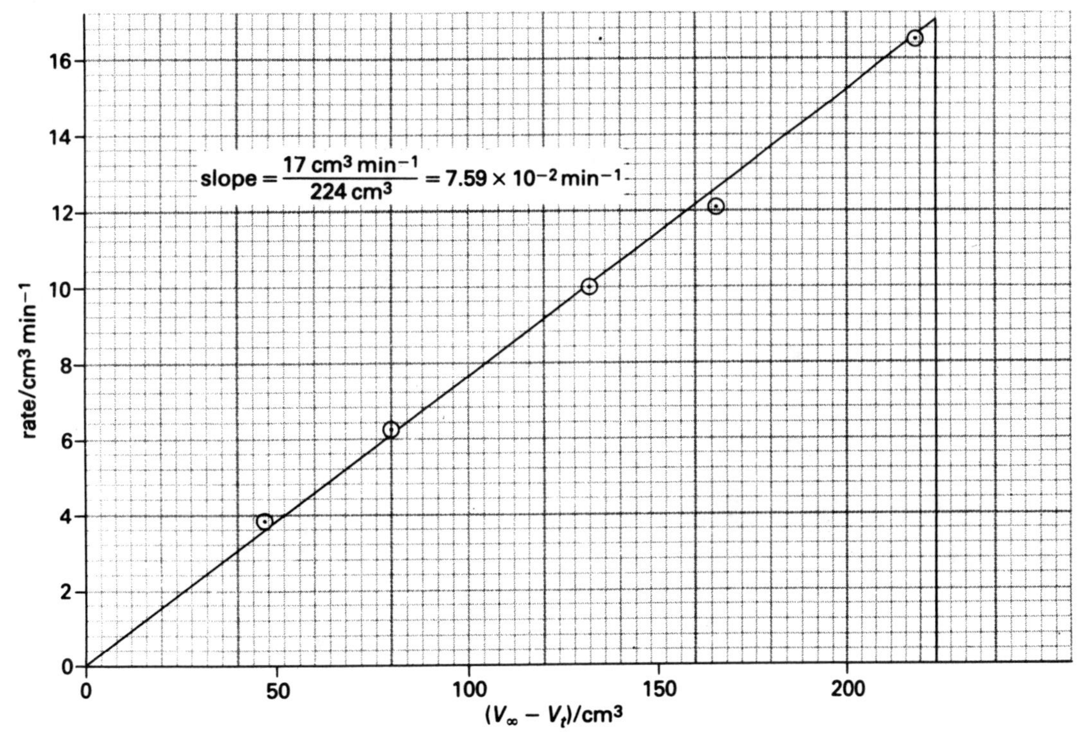

Experiment 40. Questions

1. Rate $= k\,[C_6H_5N_2{}^+Cl^-]$.

2. $k = 7.59 \times 10^{-2}$ min^{-1} at 47 °C.

3. The initial gas escape can be neglected because we are interested in the volume change, $(V_\infty - V_t)$, e.g. if you had carried out the experiment and found that $V_\infty = 150$ cm^3 and $V_t = 50$ cm^3 then $(V_\infty - V_t) = 100$ cm^3.

    Suppose the volume of gas produced in the first four minutes was 10 cm^3, then $V_\infty = 160$ cm^3 and $V_t = 60$ cm^3 but $(V_\infty - V_t)$ is still 100 cm^3. Thus we can neglect the gas produced in the first four minutes.

Experiment 41. Specimen results

Results Table 41a

Volume of 0.02 M I_2 solution/cm^3	Volume of distilled water/cm^3	$[I_2(aq)]$ /10^{-3} mol dm^{-3}	Meter reading ~~(% absorbance)~~ (% transmission)
0.0	10.0	0.0	100.0
1.0	9.0	2.0	79.5
2.0	8.0	4.0	61.0
3.0	7.0	6.0	51.0
4.0	6.0	8.0	42.0
5.0	5.0	10.0	37.0

(Continued on next page.)

(Continued from last page.)

Results Table 41c

t = time/min		% = % transmission				$I = [I_2(aq)]/10^{-3}$ mol dm^{-3}								
	t	0	$\frac{1}{2}$	1	$1\frac{1}{2}$	2	$2\frac{1}{2}$	3	$3\frac{1}{2}$	4	$4\frac{1}{2}$	5	$5\frac{1}{2}$	6
a	%	–	62.5	63.8	64.3	65.0	66.0	66.8	67.8	68.8	69.8	71.0	72.0	73.0
	I	4.00	3.90	3.72	3.66	3.56	3.44	3.32	3.20	3.07	2.94	2.80	2.70	2.58
b	%	–	62.5	64.0	65.8	67.8	69.8	71.8	74.2	76.5	79.8	82.5	85.8	89.0
	I	4.00	3.90	3.70	3.46	3.20	2.94	2.72	2.42	2.18	1.83	1.55	1.25	0.95
c	%	–	65.0	67.0	69.8	72.5	76.0	79.8	83.8	88.0	93.0	97.5	——	——
	I	4.00	3.55	3.30	2.94	2.62	2.24	1.82	1.43	1.04	0.59	0.21	——	——
d	%	–	42.5	42.6	42.8	43.5	43.6	44.2	45.0	45.3	46.0	47.0	47.5	48.5
	I	8.00	7.75	7.70	7.59	7.40	7.37	7.18	6.97	6.88	6.75	6.50	6.40	6.20
e	%	–	79.0	80.0	80.5	81.9	83.0	84.1	85.5	87.0	88.5	90.0	91.8	93.5
	I	2.00	1.92	1.82	1.76	1.62	1.51	1.40	1.26	1.13	0.97	0.85	0.70	0.55
f	%	–	63.0	64.5	65.7	67.5	69.2	70.6	72.5	75.0	77.0	79.0	81.3	84.3
	I	4.00	3.82	3.63	3.46	3.23	3.02	2.84	2.62	2.34	2.15	1.92	1.68	1.38
g	%	–	63.4	65.8	68.3	71.5	74.5	78.0	81.9	85.0	88.3	93.3	98.0	——
	I	4.00	3.75	3.48	3.13	2.75	2.40	2.02	1.63	1.32	1.01	0.58	0.17	——

<u>Initial rates of reaction</u> are given by the slopes (with sign reversed) at time zero of the curves shown on the next page.

(a) $\dfrac{(4.0 - 2.8) \times 10^{-3} \text{ mol dm}^{-3}}{6.0 \text{ min}}$ = 2.0×10^{-4} mol dm^{-3} min^{-1}

(b) $\dfrac{(4.0 - 1.8) \times 10^{-3} \text{ mol dm}^{-3}}{6.0 \text{ min}}$ = 3.7×10^{-4} mol dm^{-3} min^{-1} $\approx 2 \times$ (a)

(c) $\dfrac{(4.0 - 0.2) \times 10^{-3} \text{ mol dm}^{-3}}{6.0 \text{ min}}$ = 6.3×10^{-4} mol dm^{-3} min^{-1} $\approx 3 \times$ (a)

(d) $\dfrac{(8.0 - 6.7) \times 10^{-3} \text{ mol dm}^{-3}}{6.0 \text{ min}}$ = 2.2×10^{-4} mol dm^{-3} min^{-1} $\approx 1 \times$ (a)

(e) $\dfrac{(2.0 - 0.9) \times 10^{-3} \text{ mol dm}^{-3}}{6.0 \text{ min}}$ = 1.8×10^{-4} mol dm^{-3} min^{-1} $\approx 1 \times$ (a)

(f) $\dfrac{(4.0 - 1.9) \times 10^{-3} \text{ mol dm}^{-3}}{6.0 \text{ min}}$ = 3.5×10^{-4} mol dm^{-3} min^{-1} $\approx 2 \times$ (a)

(g) $\dfrac{(4.0 - 0.4) \times 10^{-3} \text{ mol dm}^{-3}}{6.0 \text{ min}}$ = 6.0×10^{-4} mol dm^{-3} min^{-1} $\approx 3 \times$ (a)

<u>Orders of reaction</u>

(a), (b) & (c) Initial values of $[I_2(aq)]$ and $[H^+(aq)]$ are constant.
Initial values of $[CH_3COCH_3(aq)]$ and initial rates are both in ratio 1:2:3.
Therefore, reaction is first order with respect to CH_3COCH_3.

(a), (d) & (e) Initial values of $[CH_3COCH_3(aq)]$ and $[H^+(aq)]$ are constant.
Initial values of $[I_2(aq)]$ vary but initial rate is constant. Therefore,
reaction is zero order with respect to I_2.

(a), (f) & (g) Initial values of $[CH_3COCH_3(aq)]$ and $[I_2(aq)]$ are constant.
Initial values of $[H^+(aq)]$ and initial rates are both in ratio 1:2:3.
Therefore, reaction is first order with respect to H^+.

Experiment 41. Specimen results

(Continued from last page.)

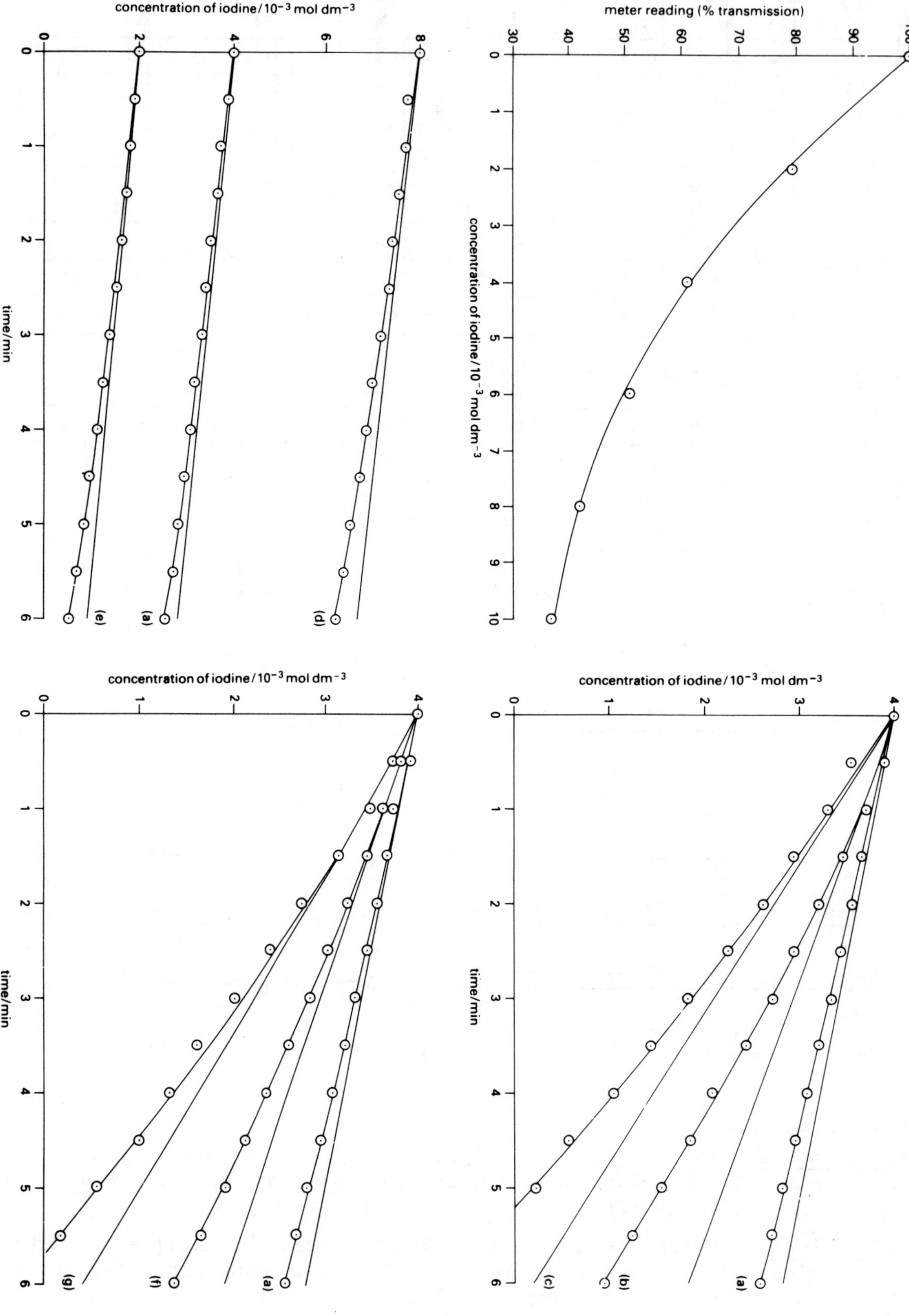

Experiment 41. Questions

1. The graphs show that the rate of reaction is directly proportional to the initial concentrations of both propanone and hydrogen ions, but is independent of the initial concentration of iodine. Thus, the reaction is first order with respect to propanone, first order with respect to hydrogen ions and zero order with respect to iodine. The rate equation is:

$$\text{rate} = k\,[CH_3COCH_3(aq)][H^+(aq)]$$

2. The overall order of reaction is two.

3. Hydrogen ions appear in the rate equation but are not reactants in the stoichiometric equation. On the other hand, iodine is a reactant but does not appear in the rate equation.

   The rate of reaction must be determined by a step in the mechanism which involves propanone and hydrogen ions but not iodine.

4. Rate constants for this reaction at different temperatures have been quoted as follows:

Temperature/$^{\circ}$C	15	20	25	30	35	40
Rate constant/10^{-3} mol^{-1} dm^3 min^{-1}	0.49	0.90	1.63	2.90	5.07	8.70

   An average value at about 23 $^{\circ}$C is obtained from our specimen results by dividing initial rates by initial concentrations as follows:

   (a) $\dfrac{2.0 \times 10^{-4}\ mol\ dm^{-3}\ min^{-1}}{(0.40 \times 0.40)\ mol^2\ dm^{-6}} = 1.25 \times 10^{-3}$

   (b) $\dfrac{3.7 \times 10^{-4}\ mol\ dm^{-3}\ min^{-1}}{(0.80 \times 0.40)\ mol^2\ dm^{-6}} = 1.16 \times 10^{-3}$

   (c) $\dfrac{6.3 \times 10^{-4}\ mol\ dm^{-3}\ min^{-1}}{(1.20 \times 0.40)\ mol^2\ dm^{-6}} = 1.31 \times 10^{-3}$

   (d) $\dfrac{2.2 \times 10^{-4}\ mol\ dm^{-3}\ min^{-1}}{(0.40 \times 0.40)\ mol^2\ dm^{-6}} = 1.38 \times 10^{-3}$

   (e) $\dfrac{1.8 \times 10^{-4}\ mol\ dm^{-3}\ min^{-1}}{(0.40 \times 0.40)\ mol^2\ dm^{-6}} = 1.13 \times 10^{-3}$

   (f) $\dfrac{3.5 \times 10^{-4}\ mol\ dm^{-3}\ min^{-1}}{(0.40 \times 0.80)\ mol^2\ dm^{-6}} = 1.09 \times 10^{-3}$

   (g) $\dfrac{6.0 \times 10^{-4}\ mol\ dm^{-3}\ min^{-1}}{(0.40 \times 1.20)\ mol^2\ dm^{-6}} = 1.25 \times 10^{-3}$

   Mean value 1.22 $\times 10^{-3}$ mol^{-1} dm^3 min^{-1}

   Your values of the rate constant may differ considerably from those we have quoted, for a number of reasons (see Q5.). Nevertheless, you should be able to deduce the rate equation without too much difficulty.

5. The main sources of error may include the following.

   (a) Temperature variation during the experiment.

   (b) Variation in light output from the bulb in the colorimeter.

   (c) Sample tubes transmitting light differently. This may be due to

       (i) different tubes having different optical characteristics,

       (ii) the same tube being in different positions relative to the bulb, and behaving as a lens,

       (iii) drops of liquid or patches of dirt on the outside of the tube.

   (d) Solutions being prepared inaccurately.

Experiment 42. Specimen results

Results Table 42

Temperature/°C	30	35	39	45	50
Temperature, T/K	303	309	312	318	323
Time, t/s	204	138	115	75	55
$\log_{10}(1/t)$	-2.31	-2.14	-2.06	-1.88	-1.74
$\frac{1}{T}$/10^{-3} K^{-1}	3.30	3.24	3.21	3.14	3.09

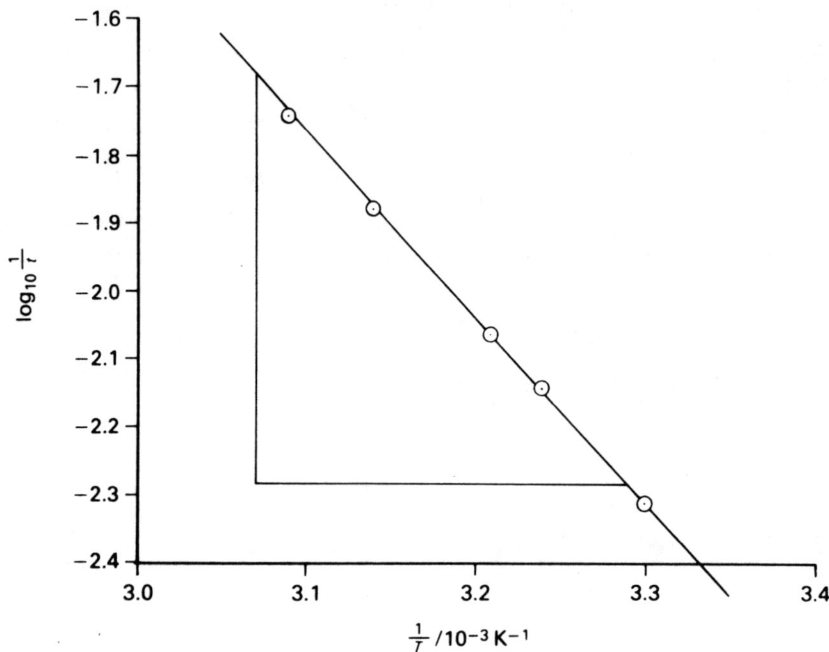

Slope of graph $= \dfrac{\Delta y}{\Delta x} = \dfrac{-2.285-(-1.675)}{(3.290-3.070) \times 10^{-3} \ K^{-1}} = -2.773 \times 10^{3} \ K$

But slope $= \dfrac{-E_a}{2.30 \ R}$

$\therefore E_a = $ -slope $\times 2.30 \ R = 2.773 \times 10^{3} \ K \times 2.30 \times 8.31 \ J \ K^{-1} \ mol^{-1}$

$= \boxed{53.0 \ kJ \ mol^{-1}}$

Experiment 43. Specimen results

Results Table 4

Temperature/°C	15	19.5	26	35	42
Temperature, T/K	288	292.5	299	308	315
Time, t/s	10.0	7.0	5.0	3.5	2.5
$\log_{10}(1/t)$	-1.00	-0.845	-0.699	-0.544	-0.398
$1/T$	3.472	3.419	3.344	3.246	3.174

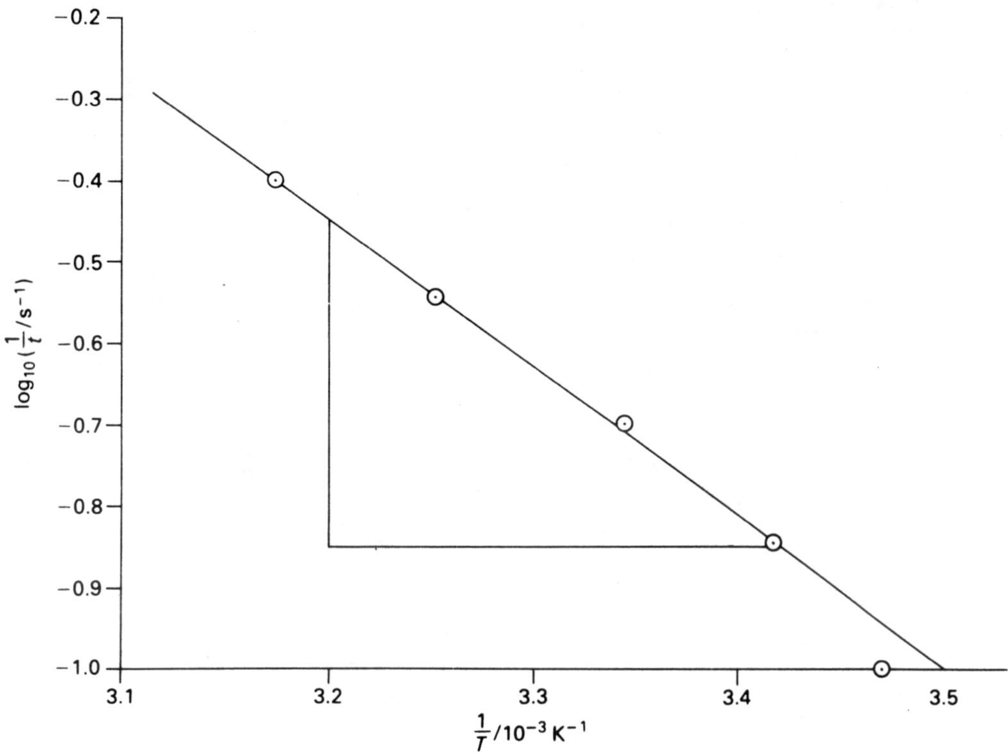

$$\text{Slope} = \frac{-0.850 - (-0.450)}{(3.420 - 3.200) \times 10^{-3}\ K} = \frac{-0.400}{0.220 \times 10^{-3}\ K} = 1820\ K$$

$$\text{But the slope} = -\frac{E_a}{2.30\ R}$$

$$\therefore\ E_a = -\text{slope} \times 2.30\ R = 1820\ K \times 2.30 \times 8.31\ J\ K\ mol^{-1} = 35\ kJ\ mol^{-1}$$

Despite the errors involved in measuring the short reaction times, this result is close to the quoted value of $28 \pm 2\ kJ\ mol^{-1}$ (School Science Review – Sept. 1975).

Experiment 44. Specimen results

Results Table 44b

Volume of $Br^-(aq)$ cm³	10.0	8.0	6.0	5.0	4.0	3.0
t/s	22.5	26.0	34.0	37.5	48.0	65.0
$\frac{1}{t}/10^{-2}$ s⁻¹	4.44	3.85	2.94	2.67	2.08	1.54
Temperature/°C	21.0	21.5	21.5	21.5	22.0	22.0

Average temperature of solutions = 21.5 °C

Results Table 44d

Volume of $BrO_3^-(aq)$/cm³	10.0	8.0	6.0	5.0	4.0	3.0
t/s	18.5	24.0	31.0	35.0	45.0	69.0
$\frac{1}{t}/10^{-2}$ s⁻¹	5.41	4.17	3.23	2.86	2.22	1.45
Temperature/°C	21.5	22.0	21.5	21.5	22.0	21.5

Average temperature of solutions = 21.5 °C

Results Table 44f

Volume of acid/cm³	10.0	8.0	6.0	5.0	4.0	3.0
t/s	13.0	19.0	30.0	39.0	60.0	97.0
$\frac{1}{t}/10^{-2}$ s⁻¹	7.69	5.26	3.33	2.56	1.67	1.03
Temperature/°C	21.5	21.5	21.5	22.0	21.5	21.5

Average temperature of solutions = 21.5 °C

Graphs 1 and 2 are straight lines (see next page).
Therefore the reaction is first order with respect to $Br^-(aq)$ and to $BrO_3^-(aq)$.
Graph 3 is not a straight line. (Note that the line must pass through the
origin since, when V = 0, t = ∞ and $1/t$ = 0.)
Plotting log $1/t$ against log (Volume of acid/cm³):

$\log\left(\frac{1}{t}/10^{-2} \text{ s}^{-1}\right)$	0.886	0.721	0.522	0.408	0.223	0.013
$\log(V/cm³)$	1.000	0.903	0.778	0.699	0.602	0.477

Slope of graph 4 = $\frac{0.60}{0.34}$ = 1.8

This suggests that the reaction is second order with respect to $H^+(aq)$.
(However, it is possible that there could be two mechanisms operating at
once, which could give a fractional order of 1.8.)

Experiment 44. Specimen results

(Continued from last page.)

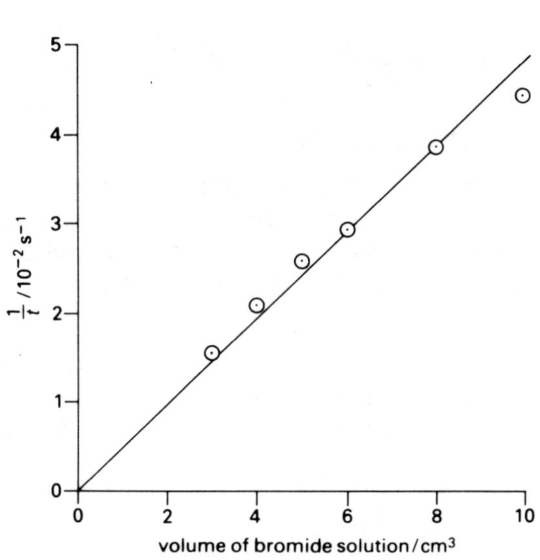

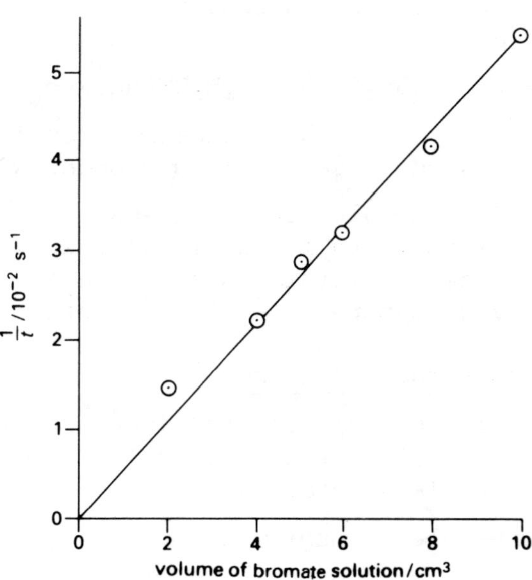

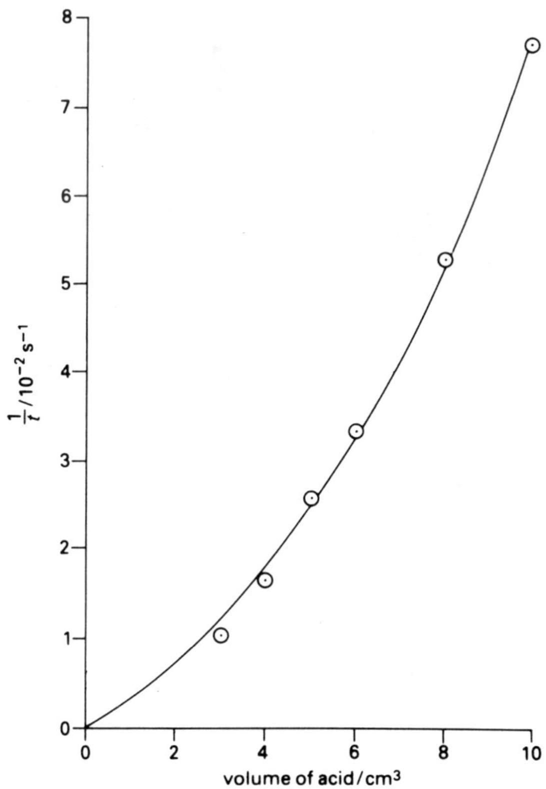

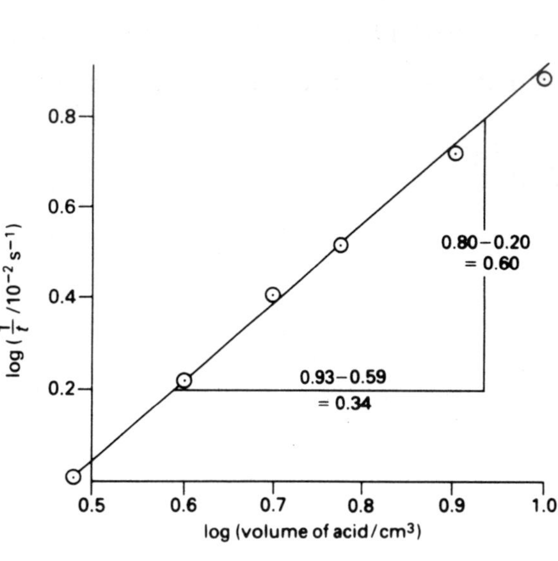

Experiment 44. Questions

1. Rate = $k[Br^-(aq)][BrO_3^-(aq)][H^+(aq)]^2$

2. The reaction producing bromine:

    $$5Br^-(aq) + BrO_3^-(aq) + 6H^+(aq) \rightarrow 3Br_2(aq) + 3H_2O(l)$$

    is slow compared with the other two. This allows the bromine to be
    removed from solution as soon as it is formed, by the reaction with
    phenol:

    $$3Br_2(aq) + C_6H_5OH(aq) \rightarrow C_6H_2Br_3OH(aq) + 3H^+(aq) + 3Br^-(aq)$$

    Similarly, the indicator reaction is fast so that as soon as the phenol
    has been used up free bromine will bleach its colour.

3. The phenol solution needs to be dilute in order to restrict the main
    reaction to its early stages, when the concentration-time curve is almost
    linear. Within this region the average rate of reaction (which is what
    you measure in a clock reaction) approximates to its initial rate.

Note on the mechanism

You may be surprised to find a reaction which is fourth order overall; this
seems to imply that four molecules collide in the rate-determining step - a
most unlikely occurrence!

A possible mechanism is:

$$H^+(aq) + Br^-(aq) \rightleftharpoons HBr(aq) \qquad \text{fast}$$

$$H^+(aq) + BrO_3^-(aq) \rightleftharpoons HBrO_3(aq) \qquad \text{fast}$$

$$HBr(aq) + HBrO_3(aq) \rightarrow \text{products} \qquad \text{slow}$$

Then rate = $k[HBr(aq)][HBrO_3(aq)]$

But $[HBr(aq)] \propto [H^+(aq)][Br^-(aq)]$ $\qquad$ (from consideration of

and $[HBrO_3(aq)] \propto [H^+(aq)][BrO_3^-(aq)]$ $\qquad$ equilibrium constants)

$\therefore$ rate = $k'[Br^-(aq)][BrO_3^-(aq)][H^+(aq)]^2$

Experiment 45. Specimen results

Results Table 45

	Observations		
Solution	Zn	Cu	Ag
$Zn^{2+}(aq)$		No change	No change
$Cu^{2+}(aq)$	The zinc became covered with a brown coating of copper. The blue solution became almost colourless.		No change
$Ag^+(aq)$	The zinc became covered with a silver-grey coat of metallic silver.	The copper became covered with a silver-grey coat of metallic silver.	

Experiment 45. Questions

1. (a) $Zn(s) + Cu^{2+}(aq) \rightarrow Zn^{2+}(aq) + Cu(s)$

    (b) $Zn(s) + 2Ag^+(aq) \rightarrow Zn^{2+}(aq) + 2Ag(s)$

    (c) $Cu(s) + 2Ag^+(aq) \rightarrow Cu^{2+}(aq) + 2Ag(s)$

2. (a) Zinc is oxidized to zinc ions $(0 \rightarrow +2)$

         Copper ions are reduced to copper $(+2 \rightarrow 0)$

    (b) Zinc is oxidized to zinc ions $(0 \rightarrow +2)$

         Silver ions are reduced to silver $(+1 \rightarrow 0)$

    (c) Copper is oxidized to copper ions $(0 \rightarrow +2)$

         Silver ions are reduced to silver $(+1 \rightarrow 0)$

3. Copper. In 1(a) copper is reduced whilst in 1(c) it is oxidized.

Experiment 46. Specimen results

Results Table 46b

Cell	Positive electrode	Negative electrode	Potential difference/V
1	Copper	Zinc	1.06
2	Silver	Copper	0.40
3	Silver	Zinc	1.48

Your readings may differ from these by up to 10% because the conditions of measurement are not standardized.

Experiment 46. Questions

1. (a) $Zn(s) \rightarrow Zn^{2+}(aq) + 2e^-$ oxidation

    When a metal goes into solution as ions, the electrons are left behind on the metal. This must be taking place to make the zinc negative.

    (b) $Cu^{2+}(aq) + 2e^- \rightarrow Cu(s)$ reduction

    When metal ions come out of solution and deposit themselves as neutral atoms, they take up electrons from the metal. This must be taking place to make the copper positive.

2. Half-reactions at the negative electrode:

    Cell 2: $Cu(s) \rightarrow Cu^{2+}(aq) + 2e^-$

    Cell 3: $Zn(s) \rightarrow Zn^{2+}(aq) + 2e^-$

    Half-reactions at the positive electrode:

    Cell 2: $2Ag^+(aq) + 2e^- \rightarrow 2Ag(s)$

    Cell 3: $2Ag^+(aq) + 2e^- \rightarrow 2Ag(s)$

3. Cell 1: $Zn(s) + Cu^{2+}(aq) \rightarrow Zn^{2+}(aq) + Cu(s)$

    Cell 2: $Cu(s) + 2Ag^+(aq) \rightarrow Cu^{2+}(aq) + 2Ag(s)$

    Cell 3: $Zn(s) + 2Ag^+(aq) \rightarrow Zn^{2+}(aq) + 2Ag(s)$

4. The reactions are identical to those which occurred in Experiment 1.

5. The salt bridge completes the circuit by allowing ions carrying charge to move between the two half-cells. These positive and negative ions replace those which are discharged at the electrodes, but contamination of the half-cell solutions is kept to a minimum.

Experiment 47. Specimen results

Results Table 47

Reaction	Additional test	Observations	Deductions
A	Add 1,1,1-trichloroethane	The lower layer became purple.	$I_2(aq)$ is present, so the reaction has taken place.
B		No change	A reaction has not taken place.
C	Add potassium hexacyano-ferrate(III)	A dark blue precipitate was formed.	$Fe^{2+}(aq)$ is present, so the reaction has taken place.
D	Add 1,1,1-trichloroethane	The lower layer became orange-brown.	$Br_2(aq)$ is present, so the reaction has taken place.
E		The purple solution became pale blue.	$MnO_4^-(aq)$ reduced to colourless $Mn^{2+}(aq)$. $Cu(s)$ oxidized to blue $Cu^{2+}(aq)$.
F		No change.	A reaction has not taken place.

Experiment 47. Question

The results do agree with predictions made using $E^{\ominus}$ values as follows:

(a) The two half-equations are:

$$Br_2 + 2e^- \rightleftharpoons 2Br^-(aq)$$

$$2I^-(aq) \rightleftharpoons I_2(aq) + 2e^-$$

which combine to make the cell:

$$Pt \mid I^-(aq), I_2(aq) \vdots Br_2(aq), Br^-(aq) \mid Pt$$

Substituting into the expression:

$$\Delta E^{\ominus} = E^{\ominus}_R - E^{\ominus}_L$$

$$\Delta E^{\ominus} = +1.07 \text{ V} - (+0.54 \text{ V}) = +0.53 \text{ V}$$

Since $\Delta E^{\ominus}$ is positive, the reaction would be expected to 'go'.

(b) The two half-equations are:

$$Br_2(aq) + 2e^- \rightleftharpoons 2Br^-(aq)$$

$$2Cl^-(aq) \rightleftharpoons Cl_2(aq)$$

which combine to make the cell:

$$Pt \mid Cl^-(aq), Cl_2(aq) \vdots Br_2(aq), Br^-(aq) \mid Pt$$

Substituting into the expression:

$$\Delta E^{\ominus} = E^{\ominus}_R - E^{\ominus}_L$$

$$\Delta E^{\ominus} = +1.07 \text{ V} - (+1.36 \text{ V}) = -0.29 \text{ V}$$

Since $\Delta E^{\ominus}$ is negative, the reaction would not 'go'.

(Continued on next page.)

(Continued from last page.)

(c) The two half-equations are:

$$Zn(s) \rightleftharpoons Zn^{2+}(aq) + 2e^-$$

$$2Fe^{3+}(aq) + 2e^- \rightleftharpoons 2Fe^{2+}(aq)$$

which combine to make the cell:

$$Zn(s) \mid Zn^{2+}(aq) \vdots Fe^{3+}(aq),Fe^{2+}(aq) \mid Pt$$

Substituting into the expression:

$$\Delta E^{\ominus} = E^{\ominus}_R - E^{\ominus}_L$$

$$\Delta E^{\ominus} = +0.77 \text{ V} - (-0.76 \text{ V}) = +1.53 \text{ V}$$

Since $\Delta E^{\ominus}$ is positive, the reaction would be expected to 'go'.

(d) The two half-equations are:

$$2MnO_4{}^-(aq) + 16H^+(aq) + 10e^- \rightleftharpoons 8H_2O(l) + 2Mn^{2+}(aq)$$

$$10Br^-(aq) \rightleftharpoons 5Br_2(aq) + 10e^-$$

which combine to make the cell:

$$Pt \mid 2Br^-(aq),Br_2(aq) \vdots [MnO_4{}^-(aq)]+ 8H^+(aq)],[4H_2O(l) + Mn^{2+}(aq)] \mid Pt$$

(Note that the stoichiometric coefficients are reduced to the smallest
whole numbers which give the correct ratio.)

Substituting into the expression:

$$\Delta E^{\ominus} = E^{\ominus}_R - E^{\ominus}_L$$

$$\Delta E^{\ominus} = +1.51 \text{ V} - (+1.07 \text{ V}) = +0.44 \text{ V}$$

Since $\Delta E^{\ominus}$ is positive, the reaction would be expected to 'go'.

(e) The two half-equations are:

$$2MnO_4{}^-(aq) + 16H^+(aq) + 10e^- \rightleftharpoons 8H_2O(l) + 2Mn^{2+}(aq)$$

$$5Cu(s) \rightleftharpoons 5Cu^{2+}(aq) + 10e^-$$

which combine to make the cell:

$$Cu(s) \mid Cu^{2+}(aq) \vdots [MnO_4{}^-(aq) + 8H^+(aq)],[4H_2O(l) + Mn^{2+}(aq)] \mid Pt$$

Substituting into the expression:

$$\Delta E^{\ominus} = E^{\ominus}_R - E^{\ominus}_L$$

$$\Delta E^{\ominus} = +1.51 \text{ V} - (+0.34 \text{ V}) = +1.17 \text{ V}$$

Since $\Delta E^{\ominus}$ is positive the reaction would be expected to 'go'.

(f) The two half-equations are:

$$3S_2O_8{}^{2-}(aq) + 6e^- \rightleftharpoons 6SO_4{}^{2-}(aq)$$

$$2Cr^{3+}(aq) + 7H_2O(l) \rightleftharpoons Cr_2O_7{}^{2-}(aq) + 14H^+(aq) + 6e^-$$

which combine to make the cell:

$$Pt \mid [2Cr^{3+}(aq) + 7H_2O(l)],[Cr_2O_7{}^{2-}(aq) + 14H^+(aq)] \vdots$$
$$S_2O_8{}^{2-}(aq),2SO_4{}^{2-}(aq) \mid Pt$$

Substituting into the expression:

$$\Delta E^{\ominus} = E^{\ominus}_R - E^{\ominus}_L$$

$$\Delta E^{\ominus} = +2.01 \text{ V} - (+1.33 \text{ V}) = +0.68 \text{ V}$$

Since $\Delta E^{\ominus}$ is positive, the reaction would be expected to 'go'.

Experiment 48. Specimen results

Results Table 48

$[\text{Ag}^+(\text{aq})]/\text{mol dm}^{-3}$	$\log\left([\text{Ag}^+(\text{aq})]/\text{mol dm}^{-3}\right)$	E/V	$E(\text{Ag}^+|\text{Ag})/\text{V}$
0.00010	-4.0	+0.25	+0.59
0.00033	-3.5	+0.28	+0.62
0.0010	-3.0	+0.31	+0.65
0.0033	-2.5	+0.34	+0.68
0.010	-2.0	+0.38	+0.72
0.10	-1.0	+0.43	+0.77

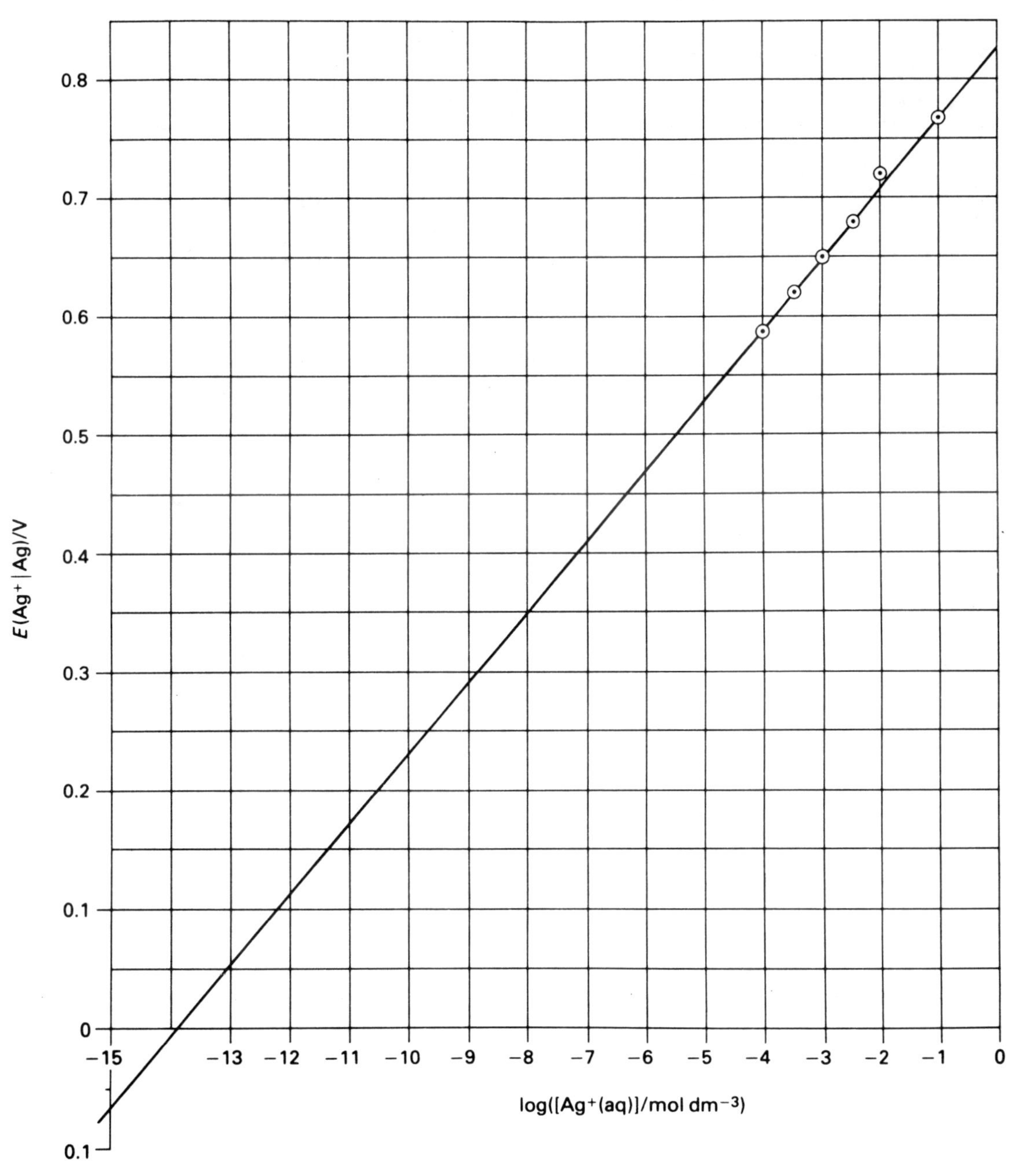

Experiment 48. Questions

1. The reaction occurring at a silver electrode is represented as:

$$Ag^+(aq) + e^- \rightleftharpoons Ag(s)$$

According to Le Chatelier's principle, an increase in $[Ag^+(aq)]$ would be expected to shift the equilibrium to the right. This would use up electrons and the silver electrode would therefore become less negative. Consequently, the electrode potential becomes less negative (more positive) as the concentration of silver ions increases.

2. The slope of the graph $= \dfrac{0.83\ V}{13.8} = 0.060\ V$

The Nernst equation is

$$E = E^\ominus + \frac{2.3\ RT}{nF}\ \log\frac{[\text{oxidized form}]}{[\text{reduced form}]}$$

which, for the silver electrode, becomes:

$$E = E^\ominus + \frac{2.3\ RT}{nF}\ \log[Ag^+(aq)]$$

The slope of the graph is therefore $\dfrac{2.3\ RT}{nF}$

Substituting values gives:

$$\frac{2.3\ RT}{nF} = \frac{2.3 \times 8.31\ J\ K^{-1}\ mol^{-1} \times 298\ K}{1 \times 96500\ C\ mol^{-1}} = 0.059\ J\ C^{-1}$$

But, by definition, $1\ J = 1\ C \times 1\ V$

$$\therefore\ 1\ V = 1\ J\ C^{-1} \quad \text{and} \quad \frac{2.3\ RT}{nF} = 0.059\ V$$

Experiment 49. Specimen results

Results Table 49b

System	$\Delta E/V$	E/V $(Ag^+|Ag)$	$\log[Ag^+(aq)]$	$[Ag^+(aq)]$ /mol dm^{-3}	$[X^-(aq)]$ /mol dm^{-3}	K_S/mol^2 dm^{-6} Calc.	Data book
1. AgCl	-0.000	+0.340	-8.20	6.3×10^{-9}	3.3×10^{-2}	2.1×10^{-10}	2.0×10^{-10}
2. AgBr	-0.135	+0.205	-10.40	4.0×10^{-11}	3.3×10^{-2}	1.3×10^{-12}	5.0×10^{-13}
3. AgI	-0.360	-0.020	-14.00	1.0×10^{-14}	3.3×10^{-2}	3.3×10^{-16}	8.0×10^{-17}
4. AgIO$_3$	+0.140	+0.480	-5.80	1.6×10^{-6}	3.3×10^{-2}	5.3×10^{-8}	2.0×10^{-8}

Calculations

1. In each case, $\Delta E = E_R - E_L = E(Ag^+|Ag) - E(Cu^{2+}|Cu)$

    $\therefore\ E(Ag^+|Ag) = \Delta E + E(Cu^{2+}|Cu) = \Delta E + 0.340$ V

    See columns 2 and 3 in Results Table 49b.

2. Values of $\log[Ag^+(aq)]$ corresponding to the values of $E(Ag^+|Ag)$ in column 3 are shown in column 4. In each case, the antilogarithm gives a value for $[Ag^+(aq)]$ shown in column 5.

3. The concentration of X^- ions at equilibrium is given by assuming a complete reaction:

    $$Ag^+(aq) + X^-(aq) \rightarrow AgX(s)$$

    10 cm^3 of 0.10 M AgNO$_3$ react with 10 cm^3 of 0.10 M KX leaving an excess of 10 cm^3 of 0.10 M KX in a total volume of 30 cm^3 of mixture.

    $\therefore\ [X^-(aq)] = 0.10$ mol dm$^{-3} \times 10$ cm^3/30 cm$^3 = 3.3 \times 10^{-2}$ mol dm^{-3}

4. $K_S = [Ag^+(aq)] \times [X^-(aq)]$

    Values of K_S in column 7 are therefore obtained by multiplying the values in columns 5 and 6.

Experiment 50. Specimen results

Results Table 50

Experiment	Observation	Conclusion
Sodium peroxide added to water plus universal indicator	Colour of gas: None. Effect on glowing splint: Relights. pH of solution: 14	The gas is oxygen.
Sodium peroxide plus water, added to acidified dichromate solution plus pentanol.	Colour of organic (top) layer is: Blue. Colour of aqueous (bottom) layer is: Very pale green.	The colour of the pentanol layer shows that a peroxide has been formed.

Experiment 50. Questions

1. (a) $Na_2O_2(s) + 2H_2O(l) \rightarrow 2NaOH(aq) + H_2O_2(aq)$

    (b) $2H_2O_2(aq) \rightarrow 2H_2O(l) + O_2(g)$

    (c) $2Na_2O_2(s) + 2H_2O(l) \rightarrow 4NaOH(aq) + O_2(g)$

2. (a) $BaO_2(s) + 2H_2O(l) \rightarrow Ba(OH)_2(aq) + H_2O_2(aq)$

    (b) (i) $BaO_2(s) + H_2SO_4(aq) \rightarrow BaSO_4(s) + H_2O_2(l)$

         (ii) It produces barium sulphate, which is insoluble and can
              be easily separated from the hydrogen peroxide by filtration.

3. (a) $Na_2O(s) + CO_2(g) \rightarrow Na_2CO_3(s)$

    (b) $2Na_2O_2(s) + 2CO_2(g) \rightarrow 2Na_2CO_3(s) + O_2(g)$

    (c) $4KO_2(s) + 2CO_2(g) \rightarrow 2K_2CO_3(s) + 3O_2(g)$

Experiment 51. Specimen results

Results Table 51a. Effect of heat on s-block carbonates

Carbonate	Time to detect CO_2	Observations
Li_2CO_3	35 s	Limewater very cloudy.
Na_2CO_3	50 s	Slight cloudiness.
K_2CO_3	70 s	Slight cloudiness.
Rb_2CO_3	not available	-
Cs_2CO_3	65 s	Slight cloudiness.
$MgCO_3$	20 s	Limewater very cloudy.
$CaCO_3$	40 s	Limewater very cloudy.
$SrCO_3$	not available	-
$BaCO_3$	no CO_2 detected	No change observed.

Results Table 51b. Effect of heat on the *s*-block nitrates

Nitrate	Time to detect O_2 or NO_2	Observations	Effect of adding dil. HCl to cold residue
$LiNO_3$	NO_2 - 35 s	Liquid formed immediately - steamy gas evolved - resolidified after 20 s	No reaction
$NaNO_3$	O_2 - 40 s	Liquid formed after 20 s. No steamy gas. Bubbles formed in liquid. Trace of brown gas observed.	Brown fumes evolved
KNO_3	O_2 - 50 s	As for $NaNO_3$	Brown fumes evolved
$RbNO_3$	not available	-	-
$CsNO_3$	O_2 - 60 s	Brown gas not detected	Brown fumes evolved
$Mg(NO_3)_2$	NO_2 - 25 s	Liquid formed immediately - steamy gas evolved - resolidified after about 20 s	No reaction
$Ca(NO_3)_2$	NO_2 - 30 s	As for Mg	No reaction
$Sr(NO_3)_2$	NO_2 - 32 s	White powder crackled - no melting	No reaction
$Ba(NO_3)_2$	NO_2 - 42 s	White powder melted after 30 s	No reaction

Experiment 51. Questions

1. Some of these nitrates are hydrated and therefore dissolve in their own water of crystallization. The water eventually boils off leaving a white solid.

2. Lithium nitrate and lithium carbonate.

3. Thermal stability of both the nitrates and carbonates increases as each group is descended.

4. Group I nitrates and carbonates are more stable.

5. (a) $4LiNO_3(s) \rightarrow 2Li_2O(s) + 4NO_2(g) + O_2(g)$

    $Li_2CO_3(s) \rightarrow Li_2O(s) + CO_2(g)$

    (b) $2KNO_3(s) \rightarrow 2KNO_2(l) + O_2(g)$. The liquid product is potassium nitrite.

    K_2CO_3 is usually reckoned to be thermally stable at Bunsen burner temperatures, although traces of CO_2 may be detected.

    (c) $2Mg(NO_3)_2(s) \rightarrow 2MgO(s) + 4NO_2(g) + O_2(s)$

    $MgCO_3(s) \rightarrow MgO(s) + CO_2(g)$

6. All Group I nitrates (except $LiNO_3$) decompose to the nitrite, which produces NO_2 (amongst other nitrogen compounds) if warmed with dilute acids. This is a useful test for the presence of nitrite ions.

Experiment 52. Specimen results

Results Table 52

Cation solution	Number of drops of anion solution added to give precipitate			
	OH^-	SO_4^{2-}	SO_3^{2-}	CO_3^{2-}
Mg^{2+}	5	40+	40+	25
Ca^{2+}	20	40+	10	8
Sr^{2+}	30 (slight)	8	2	4
Ba^{2+}	40+	1	1	2

Students' results will probably differ from ours in the numbers
of drops used, but the trends should be fairly clear.

Experiment 52. Questions

(a) The solubility of the hydroxides increases down Group II.

(b) The solubility of the sulphates decreases down Group II.

(c) The solubility of the sulphites decreases down Group II.

(d) The solubility of the carbonates decreases down Group II.

Experiment 53. Specimen results

Results Table 53

	Nature of each layer	Chlorine water	Bromine water	Iodine solution
Colour of aqueous solution.		faint green	orange-brown	reddish brown
Colour of each layer after shaking with 1,1,1-trichloroethane	Upper layer *organic/aqueous	no colour	yellow	light brown
	Lower layer *organic/aqueous	no colour	red-brown	violet
Colour of each layer after shaking with ethoxyethane	Upper layer *organic/aqueous	no colour	orange-brown	light brown
	Lower layer *organic/aqueous	no colour	yellow	yellow
Colour of each layer after shaking with hexane.	Upper layer *organic/aqueous	no colour	orange-brown	violet
	Lower layer *organic/aqueous	no colour	yellow	light brown

Experiment 53. Questions

1. The halogens are more soluble in organic solvents than water because if
    dilute aqueous solutions of the coloured halogens are mixed with organic
    solvents, most of the colour is transferred to the organic solvent.

2. Iodine has a significantly different colour in the organic layer than
    in the aqueous layer. Iodine in hexane is purple (or violet). Iodine
    in 1,1,1-trichloroethane is also purple when dilute, but red when more
    concentrated.

3. To distinguish between a dilute solution of iodine and a more concen-
    trated solution of bromine, add a few cm^3 of hexane to each solution
    and shake. The iodine appears purple in hexane while the bromine
    appears brown.

Experiment 54. Specimen results

Results Table 54

Aqueous halogen	Original colour	Colour after adding NaOH(aq)	Colour after adding H_2SO_4(aq)
Bromine water	orange-brown	very pale yellow	orange-brown
Iodine solution	reddish-brown	very pale yellow	reddish-brown

Experiment 54. Questions

1. $3Br_2(aq) + 6OH^-(aq) \rightarrow 5Br^-(aq) + BrO_3^-(aq) + 3H_2O(l)$

    $3I_2(s) + 6OH^-(aq) \rightarrow 5I^-(aq) + IO_3^-(aq) + 3H_2O(l)$

2. Both of the above reactions are disproportionation reactions.

3. Both reactions are reversible because the halogen colour returns in both
    cases when the colourless halogen-alkali mixtures are acidified.

4. Chlorine reacts with cold alkali to form chlorate(I) ion, ClO^-.

    $Cl_2(aq) + 2OH^-(aq) \rightarrow Cl^-(aq) + ClO^-(aq) + H_2O(l)$

    ClO^-(aq) is stable at room temperature whereas BrO^- and IO^- are not.

Experiment 55. Specimen results

Results Table 55

			Chlorine water	Bromine water	Iodine solution
1	Initial colour		almost none	orange	brown
2	Colour after shaking with KI		brown	brown	
	Colour of each layer after shaking with hexane	Upper	yellow	yellow	
		Lower	violet	violet	
	Conclusion		I$^-$ oxidized	I$^-$ oxidized	
3	Colour after shaking with KBr		orange		unchanged
	Colour of each layer after shaking with hexane	Upper	almost none		yellow
		Lower	red		violet
	Conclusion		Br$^-$ oxidized		no reaction
4	Colour after shaking with KCl			same	same
	Colour of each layer after shaking with hexane	Upper		red	yellow
		Lower		almost none	violet
	Conclusion			no reaction	no reaction

Experiment 55. Questions

1. (a) $I_2(KI(aq))$ does not oxidize $Cl^-(aq)$ or $Br^-(aq)$

   (b) $Br_2(aq)$ oxidizes $I^-(aq)$ but not $Cl^-(aq)$

   (c) $Cl_2(aq)$ oxidizes both $Br^-(aq)$ and $I^-(aq)$

2. $Cl_2(aq) + 2Br^-(aq) \rightarrow 2Cl^-(aq) + Br_2(aq)$

   $Cl_2(aq) + 2I^-(aq) \rightarrow 2Cl^-(aq) + I_2(aq)$

   $Br_2(aq) + 2I^-(aq) \rightarrow 2Br^-(aq) + I_2(aq)$

Experiment 56. Specimen results

Results Table 56

	Chloride	Bromide	Iodide
Action of MnO_2 and conc. H_2SO_4	Steamy fumes in cold. Pale green gas on warming. Choking smell.	Steamy brown fumes in cold. More on warming. Choking smell.	Steamy violet fumes in cold. More on warming. Bad egg smell.
Suspected product	$Cl_2(HCl)$	$Br_2(HBr)$	$I_2(HI, H_2S)$
Confirmatory test	Blue litmus bleached. Starch/ iodide paper blue.	Red colour when bubbled into hexane.	Violet colour in hexane.
Action of conc. H_2SO_4	Steamy fumes in cold. No green gas even on heating.	Steamy fumes in cold. Brown gas on warming.	Steamy violet fumes in cold. More on warming. Bad egg smell.
Suspected product	HCl	$HBr(Br_2)$	$I_2(HI, H_2S)$
Confirmatory test	White smoke with ammonia. Blue litmus turns red but not bleached.	White smoke with ammonia. Red colour in hexane.	Violet colour in hexane.
Action of H_3PO_4	Steamy fumes on warming.	Steamy fumes on warming.	Steamy fumes on warming.
Suspected product	HCl	HBr	HI
Confirmatory test	White smoke with ammonia.	White smoke with ammonia.	White smoke with ammonia.

Experiment 56. Questions

1. (a) $KCl(s) + H_2SO_4(l) \rightarrow KHSO_4(s) + HCl(g)$ (K_2SO_4 at high temperature only)

   (b) $2KCl(s) + H_3PO_4(l) \rightarrow K_2HPO_4(s) + 2HCl(g)$

   (c) $KBr(s) + H_2SO_4(l) \rightarrow HBr(g) + KHSO_4(s)$

   $2HBr(g) + H_2SO_4(l) \rightarrow SO_2(g) + 2H_2O(l) + Br_2(l)$

   (d) $2KBr(s) + H_3PO_4(l) \rightarrow K_2HPO_4(s) + 2HBr(g)$

   (e) $KI(s) + H_2SO_4(l) \rightarrow KHSO_4(s) + HI(g)$

   $2HI(g) + H_2SO_4(l) \rightarrow SO_2(g) + 2H_2O(l) + I_2(s)$ or

   $8HI(g) + H_2SO_4(l) \rightarrow H_2S(g) + 4I_2(s) + 4H_2O(l)$

   (f) $2KI(s) + H_3PO_4(l) \rightarrow K_2HPO_4(s) + 2HI(g)$

2. Concentrated H_2SO_4 alone is not a sufficiently strong oxidizing agent to oxidize HCl(g) to $Cl_2(g)$. This is due to the strong H—Cl bond. The combination of concentrated H_2SO_4 and MnO_2 can oxidize HCl(g) to $Cl_2(g)$.

3. Concentrated H_2SO_4 alone is a sufficiently strong oxidizing agent to oxidize HI to I_2. This is due to the weakness of the H—I bond. A more powerful oxidizing agent such as concentrated H_2SO_4 and MnO_2 will merely produce similar results.

Experiment 57. Specimen results·

Results Table 57

Test	Chloride	Bromide	Iodide
Action of AgNO$_3$(aq) Effect of standing in (a) dark (b) light	White ppt. No change. Darkens rapidly.	Pale yellow ppt. No change. Darkens slightly.	Yellow ppt. No change. No change.
Action of AgNO$_3$(aq) followed by dilute HNO$_3$(aq)	White ppt. insoluble in HNO$_3$(aq).	Pale yellow ppt. insoluble in HNO$_3$(aq).	Yellow ppt. insoluble in HNO$_3$(aq).
Action of AgNO$_3$(aq) followed by NH$_3$(aq)	White ppt. soluble in NH$_3$(aq).	Pale yellow ppt. slightly soluble in NH$_3$(aq).	Yellow ppt. insoluble in NH$_3$(aq).
Action of Pb(NO$_3$)$_2$(aq)	White ppt.	White ppt.	Yellow ppt.
Action of H$_2$O$_2$(aq) and dilute H$_2$SO$_4$(aq)	No change.	No change.	Brown colour. Violet in hexane

Experiment 57. Questions

1. (a) $Ag^+(aq) + Cl^-(aq) \rightarrow AgCl(s)$

     $Ag^+(aq) + Br^-(aq) \rightarrow AgBr(s)$

     $Ag^+(aq) + I^-(aq) \rightarrow AgI(s)$

   (b) $Pb^{2+}(aq) + 2Cl^-(aq) \rightarrow PbCl_2(s)$

     $Pb^{2+}(aq) + 2Br^-(aq) \rightarrow PbBr_2(s)$

     $Pb^{2+}(aq) + 2I^-(aq) \rightarrow PbI_2(s)$

2. (a) Add silver nitrate solution to each solution in turn, followed by dilute or concentrated ammonia solution. A white precipitate is obtained with the chloride ion which is soluble in dilute ammonia solution. A pale yellow precipitate is obtained with bromide ions which is slightly soluble in dilute ammonia solution and soluble in concentrated ammonia solution.

   (b) Add lead ions to bromide and iodide solutions separately. A white precipitate is obtained with the Br$^-$ ions and a yellow precipitate with the I$^-$ ions.

3. (a) $2I^-(aq) + H_2O_2(aq) + 2H^+(aq) \rightarrow I_2(aq) + 2H_2O(l)$

   (b) H_2O_2 is not a sufficiently strong oxidizing agent to oxidize Cl$^-$(aq) and Br$^-$(aq) ions.

4. The light converts some of the silver halide to small particles of metallic silver, which darkens the precipitates.

Experiment 58. Specimen results

Results Table 58

	NaClO	NaClO$_3$	NaIO$_3$
Effect of heat on solid		Oxygen given off (splint relit). Colourless residue.	Oxygen given off (splint relit). Colourless residue. (Further changes on prolonged heating.
pH of solution	13	7	7
Addition of neutral KI(aq)	Iodine formed as brown colouration.	No change.	Slight yellow colour after 30 s (trace of iodine).
Addition of acidified KI(aq)	Iodine formed as brown colouration.	Slight yellow colour after 30 s (trace of iodine).	Iodine formed as brown colouration.
Addition of Co^{2+}(aq)	Oxygen given off (splint relit). Black ppt. (CoO)	No change.	No change.
Addition of dilute H$_2$SO$_4$(aq)	Chlorine given off (litmus bleached).	No change.	No change.

Experiment 58. Questions

1. (a) $4NaClO_3(s) \rightarrow NaCl(s) + 3NaClO_4(s)$ (below 400 $^\circ$C) and

   $NaClO_4(s) \rightarrow NaCl(s) + 2O_2(g)$ (above 400 $^\circ$C)

   $2NaIO_3(s) \rightarrow 2NaI(s) + 3O_2(g)$

2. The common product is iodine. Each salt oxidizes acidified KI to iodine, I_2

3. NaClO(aq) is the strongest oxidizing agent; only this salt oxidizes neutral KI solution.

4. KBrO$_3$(aq) will oxidize acidified KI to iodine.

5. NaClO$_4$ is a relatively weak oxidizing agent.

6. $2ClO^-(aq) \rightarrow 2Cl^-(aq) + O_2(aq)$

7. $ClO^-(aq) + 2H^+(aq) + Cl^-(aq) \rightarrow Cl_2(g) + H_2O(l)$

Experiment 59. Specimen results

Results Table 59

Solution in flask	iodide/iodate/acid mixture		
Solution in burette	sodium thiosulphate	0.10	mol dm^{-3}
Indicator	starch		

		Trial	1	2	3	4
Burette readings	Final		30.4	30.30	30.20	30.40	-
	Initial		0.0	0.00	0.00	0.20	-
Volume used/cm^3			30.4	30.30	30.20	30.20	-
Mean titre/cm^3	30.20					

Experiment 59. Calculations

1. $n = cV = 0.10$ mol dm^{-3} $\times$ $\dfrac{30.2}{1000}$ dm^3 $= 30.2 \times 10^{-4}$ mol

2. Amount of iodine atoms = amount of thiosulphate = 30.2×10^{-4} mol

3. $n = cV = 0.10$ mol dm^{-3} $\times$ $\dfrac{5.0}{1000}$ dm^3 $= 5.0 \times 10^{-4}$ mol

4. Amount of I from KI = total amount − amount from KIO$_3$

$$= (30.2 - 5.0) \times 10^{-4} \text{ mol}$$

$$= 25.2 \times 10^{-4} \text{ mol}$$

5. 5.0×10^{-4} mol of KIO$_3$ reacts with 25.2×10^{-4} mol of KI

$\therefore$ 1 mol of KIO$_3$ reacts with $\dfrac{25.2}{5.0}$ mol of KI = 5 mol

This is in agreement with the equation:

$$IO_3^-(aq) + 6H^+(aq) + 5I^-(aq) \rightarrow 3I_2(aq) + 3H_2O(l)$$

Experiment 60 Specimen results

Results Table 60a

Test	Observations	Inferences
(a) Heat approximately 0.1 g of F in a pyrex tube, at first gently and then more strongly, until the change is complete. Cool and keep the residue. Test any gases evolved.	White solid decrepitates on heating giving first a liquid, then a white solid. A colourless gas is given off which relights a glowing splint. On further heating, the white solid melts to a clear liquid which eventually turns orange. On cooling, the orange liquid becomes colourless and finally solidifies to a white solid.	O_2 evolved. Orange solution suggests the formation of Br_2. F could be BrO_3^- or IO_3^-. The final residue is probably an alkali metal bromide or iodide.
(b) Make an aqueous solution of the residue from (a) and carry out the following tests on portions:		
(i) Add aqueous silver nitrate followed by dilute nitric acid.	A cream precipitate forms which is insoluble in dilute HNO_3.	AgBr (or possibly AgI) precipitated from Br^- (or I^-) in residue.
(ii) Add aqueous chlorine.	A yellow solution is produced.	Br_2 (or possibly I_2) displaced by Cl_2 from halide.
(iii) Add aqueous lead(II) ethanoate (lead acetate).	A white precipitate is formed.	Confirms Br^- not I^- in the residue (PbI_2 is yellow).

Results Table 60b

Substance provided	Name of non-metal	Name and formula of substance/ion	Oxidation number
F	Bromine	Bromate BrO_3^- Bromide Br^-	+V (+5) -I (-1)

Experiment 61. Specimen results

Results Table 61a   Tests on unknown substance D

Method	Observations	Inferences
(a) A few crystals of D were heated in an ignition tube, gently at first and then more strongly.	Some decrepitation at first. Solid eventually melted to a clear liquid which turned brown then red on further heating and gave off a little purple vapour. Cooled to a colourless solid.	Purple vapour must be iodine, probably from thermal decomposition of potassium iodide.
(b) The residue from (a) was cooled and one drop of concentrated H_2SO_4 was added. Then a few more drops were added.	The mixture immediately darkened (almost black) and steamy fumes with a choking (bad egg) smell were released. A brownish liquid condensed on the tube just above the mixture.	HI produced (displaced from iodide by non-volatile H_2SO_4). Some oxidation of HI to iodine by concentrated H_2SO_4 which was reduced to H_2S.
(c) Ten drops diluted H_2SO_4 were added to 5 drops $KMnO_4$ solution followed by a few crystals of D. The mixture was then warmed.	The solution immediately changed from purple to dark brown. No further change on warming.	I^- oxidized to I_2 (brown in solution as I_3^-)

Results Table 61c   Experiments to test inferences

Inference tested	Test and observations	Conclusion
Iodine produced by reaction of D and E in acid solution.	About 1 cm^3 of hexane was added to mixture of D, E and acid ((e) above). The tube was shaken gently. The upper layer of hexane turned purple.	Iodine confirmed
D is potassium iodide.	A few crystals of D were dissolved in distilled water and a few drops of dilute HNO_3 were added followed by $AgNO_3$ solution. Yellow precipitate formed.	Yellow precipitate of AgI confirms that D is potassium iodide.

<u>Final conclusion.</u> D is potassium iodide, KI, and E is potassium iodate(V), KIO_3

Results Table 61b · Tests on unknown substance E

Method	Observations	Inferences
(a) A few crystals of E were heated in an ignition tube, gently at first and then more strongly.	No change at first, but solid soon melted to colourless liquid and effervesced. Gas was colourless at first, then tinged with purple. Red liquid eventually formed on strong heating which cooled to a colourless solid.	Purple vapour must be iodine but E is not the same as D. Perhaps D is produced from E by decomposition.
Gas was tested with damp litmus papers, blue and red.	No colour change.	Neutral gas (or CO_2) produced.
Gas was drawn into teat pipette and bubbled through a little lime-water.	No change.	No CO_2 produced.
A glowing splint was lowered into the tube.	The splint relit.	O_2 produced, probably by decomposition of potassium iodate(V).
(b) The residue from (a) was cooled and one drop of conc. H_2SO_4 was added. Then a few more drops were added.	The mixture immediately turned almost black and colourless fumes with a choking 'bad egg' smell were released. A brownish liquid condensed on the tube above the mixture.	HI produced (displaced from iodide by non-volatile H_2SO_4). Some oxidation of HI to iodine by concentrated H_2SO_4 which was reduced to H_2S. This suggests that the residue contains I^-, thus E is very likely to be KIO_3.
(c) 10 drops of dilute H_2SO_4 were added to 5 drops of $KMnO_4$ solution followed by a few crystals of E. The mixture was then warmed.	No change observed apart from dissolving of crystals.	No redox reaction occurs. This is consistent with the identity suggested for E, i.e. KIO_3.
(d) A few crystals of D and E were dissolved separately in about 2 cm^3 of distilled water in each of two tubes. The solutions were then mixed.	D dissolved very easily, E less easily. Both solutions were colourless and remained so on mixing (perhaps a slight yellow colouration).	No apparent reaction. This again is consistent for KI(aq) and KIO_3(aq) in neutral solution.
(e) Dilute H_2SO_4 was added dropwise to the mixed solutions from (d)	Solution turned dark brown with some blackish precipitate.	I_2 formed by redox reaction between I^- and IO_3^- in presence of H^+.

Experiment 63. Specimen results

Results Table 62a

	NaCl	$MgCl_2$	$AlCl_3$	$SiCl_4$	PCl_3	S_2Cl_2
Appearance	Colourless solid	Colourless solid	Colourless solid	Colourless liquid'	Colourless liquid	Yellow liquid
On mixing with water		(Hydrated)	(Anhydrous)			
Initial temp.	17 °C	17 °C	17 °C	17 °C	17 °C	17 °C
Final temp.	17 °C	19 °C	49 °C	19 °C	27 °C	20 °C
Does it dissolve?	Yes	Yes	Yes	Yes	Yes (eventually)	Yes
pH of solution	7	6	4 or less	Below 4	Below 4	Below 4
Other observations (if any)	None	None	Vigorous reaction. Acid gas— white fumes with NH_3.	Acid gas— white fumes with NH_3. White particles in liquid.	Acid gas— white fumes with NH_3.	Cloudy solution formed.
On mixing with hexane						
Initial temp.	16 °C	16 °C	16 °C	16 °C	17 °C	17 °C
Final temp.	16 °C	16 °C	16 °C	16 °C	17 °C	17 °C
Does it dissolve?	No	No	Very slightly.	Slightly soluble if at all.	Yes	Yes
Other observations (if any)	None	None	None	Two layers.	None	None

Experiment 62. Questions

1.

Table 62b. Properties of Chlorides of Period 3

Formula of chloride	NaCl	$MgCl_2$	$AlCl_3$	$SiCl_4$	PCl_3	S_2Cl_2	Cl_2
Melting-point/°C	80	714	190	-70	-112	-80	-101
Boiling-point/°C	1467	1412	183	58	76	136	-34.6
Physical state at r.t.p.*	Solid	Solid	Solid	Liquid	Liquid	Liquid	Gas
$\Delta H^{\ominus}_f$/kJ mol^{-1}	-411	-642	-695	-640	-339	-60.2	0
$\Delta H^{\ominus}_f$ per mole of Cl/kJ mol^{-1}	-411	-321	-232	-160	-113	-30.1	0
Conductivity of liquid	Good	Good	Poor	Very poor	Very poor	Very poor	Very poor
Action of water	Dissolves	Dissolves	Reacts	Reacts	Reacts	Reacts	Reacts
pH of aqueous solution	7	6.5	3	2	2	2	2
Solubility in hexane	Insoluble	Insoluble	Insoluble	Dissolves	Dissolves	Dissolves	Dissolves
Structure	Giant	Giant	Simple molecular**	Simple molecular	Simple molecular	Simple molecular	Simple molecular
Bonding	Ionic	Ionic	Covalent	Covalent	Covalent	Covalent	Covalent

*r.t.p. = room temperature and pressure (i.e. 20 °C and 1 atm)
**The solid has a layer structure. The vapour contains Al_2Cl_6 molecules.

(Continued on next page.)

(Continued from last page.)

2. (a) $AlCl_3(s) + 6H_2O(l) \rightarrow [Al(H_2O)_6]^{3+}(aq) + 3Cl^-(aq)$
                        hexaaquaaluminium(III) ion

The $[Al(H_2O)_6]^{3+}$ ions dissociate to form H_3O^+ ions giving rise to
an acidic solution. There is also some HCl gas produced.

   (b) $SiCl_4(l) + (x + 2)H_2O(l) \rightarrow SiO_2 \cdot xH_2O(aq) + 4HCl(g)$
                        hydrated silicon(IV) oxide

(Some books refer to hydrated silicon(IV) oxide as silicic acid
and give it the formula H_2SiO_3 or $Si(OH)_4$. These species may exist
in solution but it seems likely that a range of hydrates such as
$SiO_2 \cdot 2H_2O$ and $SiO_2 \cdot H_2O$ exist which we represent as $SiO_2 \cdot xH_2O$. The
hydrated oxide appears as a colloidal solution or a gel.)

   (c) $PCl_3(l) + 3H_2O(l) \rightarrow H_2PHO_3(aq) + 3HCl(aq)$
                        phosphonic acid

Phosphonic acid H_2PHO_3 is sometimes called phosphorous acid and
given the formula H_3PO_3.

Experiment 63. Specimen results

Results Table 63

Mass of aluminium	0.25 g
Mass of empty specimen tube, m_1	10.20 g
Mass of specimen tube and product, m_2	10.60 g
Mass of product, $m = (m_2 - m_1)$	0.40 g
% yield	33%

Experiment 63. Questions

1. Any moisture present in the apparatus, such as water vapour mixed with the chlorine, would hydrolyse the aluminium chloride as it formed.

2. The product would contain aluminium oxide as an impurity. It would be formed by the reaction

$$2AlCl_3(s) + 3H_2O(g) \rightarrow Al_2O_3(s) + 6HCl(g)$$

3. Using the data from the specimen results the calculation is as follows.

$$Al(s) + 1\tfrac{1}{2}Cl_2(g) \rightarrow AlCl_3(s)$$

First, calculate the mass of $AlCl_3$ that would theoretically be formed from 0.25 g of Al

27 g of Al produces   133.5 g of $AlCl_3$

1.0 g of Al produces $\dfrac{133.5 \text{ g}}{27}$ of $AlCl_3$

0.25 g of Al produces $\dfrac{133.5 \text{ g}}{27} \times 0.25 = 1.2$ g of $AlCl_3$

$$\% \text{ yield} = \dfrac{\text{actual mass of product}}{\text{theoretical mass of product}} \times 100$$

$$\therefore \% \text{ yield} = \dfrac{0.40 \text{ g}}{1.2 \text{ g}} \times 100 = \boxed{33\%}$$

This experiment often gives low yields. Likely reasons are:
(a) Some product remains in combustion tube and receiver bottle.
(b) Some product is carried away with excess chlorine.
(c) Some aluminium reacts with residual air to form oxide which remains in the combustion tube.
(d) Some product is hydrolysed by residual water vapour.

4. A desiccator is necessary to prevent hydrolysis by moist air.

Experiment 64. Specimen results

Results Table 64a

	Na_2O_2	MgO	Al_2O_3	SiO_2	P_4O_{10}	SO_2
Appearance	White or pale yellow solid	White solid	White solid	White solid	White solid	Colourless gas
On mixing with water						
Initial temperature	22 °C	22 °C	23 °C	23 °C	22 °C	22 °C
Final temperature	24 °C	23 °C	23 °C	23 °C	25 °C	23 °C
Does it dissolve?	Yes	Partially	No	No	Yes	Yes
pH of solution	14	12	7	7	below 4	below 4
Other observation(s) (if any)	Vigorous reaction. Colourless gas given off. Relights glowing splint.	None	None	None	Vigorous reaction, steamy fumes evolved.	

Experiment 64. Questions

1.

Table 64b

Formula of oxide	Na_2O_2*	MgO	Al_2O_3	SiO_2	P_4O_{10}*	SO_2*	Cl_2O*
Melting-point/°C	460	2900	2040	1610(quartz)	580	-75	-20
Boiling-point/°C	Decomposes	3600	2980	2230(quartz)	300	-10	2
State at s.t.p.	Solid	Solid	Solid	Solid	Solid	Gas	Gas
Action of water	Reacts	None apparent	None apparent	Insoluble	Reacts	Reacts	Reacts
pH of aq. solution	11-12	8-9	7-8	6-7	Below 3	3 or less	3 or less
Acid/base nature	Basic	Basic	Amphoteric	Acidic	Acidic	Acidic	Acidic
Conductivity of liquid	Good	Good	Good	Very poor	Very poor	Very poor	Very poor
Solubility in hexane	Insoluble	Insoluble	Insoluble	Insoluble	Slightly soluble	Dissolves	Dissolves
Structure	Giant	Giant	Giant	Giant	Simple molecular	Simple molecular	Simple molecular
Bonding	Ionic	Ionic	Ionic	Covalent	Covalent	Covalent	Covalent

Those substances marked * represent the most familiar or readily available oxides of that element.

2. (a) $Na_2O_2(s) + 2H_2O(l) \rightarrow 2NaOH(aq) + H_2O_2(aq)$

   $2H_2O_2(aq) \rightarrow 2H_2O(l) + O_2(g)$

   (b) Magnesium oxide reacts slightly in cold water and moderately in hot to form the hydroxide, which is very slightly soluble.

   $MgO(s) + H_2O(l) \rightarrow Mg(OH)_2(s)$

   (c) $P_4O_{10}(s) + 6H_2O(l) \rightarrow 4H_3PO_4(aq)$

   (d) $SO_2(g) + H_2O(l) \rightarrow H_2SO_3(aq)$

   (e) $Cl_2O(g) + H_2O(l) \rightarrow 2HClO(aq)$

3. The oxides change from being giant ionic structures in Groups I, II and III through a giant covalent structure for silicon(IV) oxide, to simple covalent molecular structures for phosphorus, sulphur and chlorine.

4. The acid-base nature changes from basic for the *s*-block oxides through amphoteric for aluminium oxide to acidic for the oxides of the remaining elements.

5. As the structure and bonding of the oxides changes from giant ionic to simple molecular across the period so their acid-base nature changes from basic to acidic.

Experiment 65. Specimen results

Results Table 65a

Acid	Tin	Lead
Dilute hydrochloric acid	Colourless gas evolved on heating which 'popped' with a lighted splint.	Slight reaction (if any) on heating.
Concentrated hydrochloric acid	Colourless gas evolved on heating which 'popped' with a lighted splint.	Colourless gas evolved on heating. 'Popped' with difficulty. White solid produced in test-tube.
Concentrated nitric acid	Brown gas evolved on heating.	Brown gas evolved on heating.

Results Table 65b

Reagent	Sn^{2+}(aq)(acidified)	Pb^{2+}(aq)
(a) Sodium hydroxide solution	White precipitate dissolved in excess sodium hydroxide solution.	White precipitate dissolved in excess sodium hydroxide solution.
(b) Ammonia solution	White precipitate insoluble in excess ammonia solution.	White precipitate insoluble in excess ammonia solution.
(c) Dilute hydrochloric acid	No reaction.	White precipitate dissolved on boiling and crystallized on cooling to form white crystals.
(d) Acidified potassium manganate(VII) solution	Decolorized	No reaction.
(e) Potassium chromate (VI) solution	Green-blue solution obtained.	Yellow precipitate formed.
(f) Sodium sulphide solution	Brown precipitate formed.	Dark brown-black precipitate formed.
(g) Potassium iodide solution	No reaction.	Yellow precipitate formed.

Experiment 65. Questions

1. (a) $Pb(s) + 2HCl(aq) \rightarrow PbCl_2(s) + H_2(g)$ ($PbCl_2(aq)$ if hot)

         $Sn(s) + 2HCl(aq) \rightarrow SnCl_2(aq) + H_2(g)$

2. Yes, metals above hydrogen in the electrochemical series produce
    hydrogen on reaction with acids.

3. (a) Nitrogen dioxide(NO_2) is responsible for the brown colour.

    (b) Yes, zinc (or copper) reacts in a similar way.

    (c) Nitric acid is a strong oxidizing agent.

4. Table 65c

Equations	Comments
$Sn^{2+}(aq) + 2OH^-(aq) \rightarrow Sn(OH)_2(s)$ $Sn(OH)_2(s) + 4OH^-(aq) \rightarrow Sn(OH)_6^{4-}(aq)$	The precipitate dissolves in excess NaOH to form a stannate(II) ion* $Sn(OH)_6^{4-}$.
$Pb^{2+}(aq) + 2OH^-(aq) \rightarrow Pb(OH)_2(s)$ $Pb(OH)_2(s) + 4OH^-(aq) \rightarrow Pb(OH)_6^{4-}(aq)$	The precipitate dissolves in excess NaOH to form a plumbate(II) ion $Pb(OH)_6^{4-}$.
$2MnO_4^-(aq) + 16H^+(aq) + 5Sn^{2+}(aq)$ $\rightarrow 2Mn^{2+}(aq) + 5Sn^{4+}(aq)*** + 8H_2O(l)$	Redox reaction. Sn^{2+} is behaving as a reducing agent.
$Pb^{2+}(aq) + 2Cl^-(aq) \rightarrow PbCl_2(s)$	Simple precipitation reaction. Precipitate soluble in hot water.
$Cr_2O_7^{2-}(aq)** + 14H^+(aq) + 3Sn^{2+}(aq)$ $\rightarrow 2Cr^{3+}(aq) + 3Sn^{4+}(aq)*** + 7H_2O(l)$	Redox reaction. Sn^{2+} is behaving as a reducing agent. Cr^{3+} gives the solution its blue-green colour.
$Pb^{2+}(aq) + CrO_4^{2-}(aq) \rightarrow PbCrO_4(s)$	Simple precipitation reaction.
$Pb^{2+}(aq) + S^{2-}(aq) \rightarrow PbS(s)$	Simple precipitation reaction.
$Sn^{2+}(aq) + S^{2-}(aq) \rightarrow SnS(s)$	Simple precipitation reaction.
$Pb^{2+}(aq) + 2I^-(aq) \rightarrow PbI_2(s)$	Simple precipitation reaction. The precipitate is soluble in hot water and recrystallizes on cooling as golden yellow platelets.

*Various formulae have been proposed for the stannate(II) ion ranging
from SnO_2^{2-} for the anhydrous form to $Sn(OH)_4^{2-}$ and $Sn(OH)_6^{4-}$ for the
hydrated forms. $Sn(OH)_6^{4-}$ seems the most probable. Similar variations
have been proposed for the plumbate(II) ion, $Pb(OH)_6^{4-}$.

**Chromate(VI) (CrO_4^{2-}) changes to dichromate(VI) ($Cr_2O_7^{2-}$) when acidified
($2CrO_4^{2-}(aq) + 2H^+(aq) \rightleftharpoons Cr_2O_7^{2-}(aq) + H_2O(l)$).

***Sn^{4+} ions are stated in the equations for convenience; you should
bear in mind that tin(IV) compounds are mainly covalent. Aqueous
solutions of tin(IV) usually contain complexes.

(Continued on next page.)

(Continued from last page.)

5. If potassium manganate(VII) solution were acidified with hydrochloric
   acid or sulphuric acid the chloride or sulphate ions would react with
   the lead ions to form a white precipitate of lead(II) chloride or
   lead(II) sulphate.

$$Pb^{2+}(aq) + 2Cl^-(aq) \rightarrow PbCl_2(s)$$

$$Pb^{2+}(aq) + SO_4^{2-}(aq) \rightarrow PbSO_4(s)$$

   Ethanoate ions will not form a precipitate with lead ions.

6. Lead(II) ions are more stable because they are not oxidized by any of the
   reagents used in the experiment. Tin(II) ions are oxidized to the +4
   state by acidified solutions of potassium manganate(VII) and potassium
   dichromate(VI).

7. The tin(II) ions are likely to be oxidized by the mercury(II) chloride
   forming the tin(IV) compound and mercury. The final appearance of the
   mixture is therefore likely to be grey due to the formation of mercury.
   This does in fact occur, although mercury(I) chloride (Hg_2Cl_2) is
   initially formed as a white solid.

$$2HgCl_2(aq) + Sn^{2+}(aq) \rightarrow Hg_2Cl_2(s) + Sn^{4+}(aq) + 2Cl^-(aq)$$

$$Hg_2Cl_2(s) + Sn^{2+}(aq) \rightarrow 2Hg(l) + Sn^{4+}(aq) + 2Cl^-(aq)$$

Experiment 66. Specimen results

Results Table 66a

Mass of Pb	5.0 g
Mass of PbO_2	3.9 g
% yield	67%
Appearance of PbO_2	Red/brown solid
Appearance of SnO_2	White/pale yellow solid

Percentage yield of PbO_2 is calculated as follows:

1 mol of Pb produces 1 mol of PbO_2

i.e. 207 g of Pb produces 239 g of PbO_2

$\therefore$ 1 g of Pb produces $\dfrac{239 \text{ g}}{207}$ of PbO_2

$\therefore$ 5.0 g of Pb produces $\dfrac{239 \text{ g}}{207} \times 5.0 = 5.8$ g of PbO_2 (theoretical maximum)

$$\% \text{ yield} = \frac{\text{actual mass of product}}{\text{theoretical mass of product}} \times 100$$

$\therefore$ % yield $= \dfrac{3.9 \text{ g}}{5.8 \text{ g}} \times 100 = \boxed{67\%}$

Results Table 66b

Test	Tin(IV) oxide	Lead(IV) oxide
(a) Heat	No change.	Red liquid formed on heating and colourless gas evolved which relit a glowing splint (O_2 evolved). Yellow glassy solid remained.
(b) Dilute hydro-chloric acid	No reaction.	Slow evolution of a gas which bleached damp litmus paper. Characteristic smell of chlorine detectable.
(c) Concentrated hydrochloric acid	No apparent reaction.	Vigorous reaction, chlorine evolved, yellow clear liquid remained with a white solid at the bottom of the test-tube.
(d) Concentrated sodium hydroxide	A very small quantity probably dissolved but this was not obvious.	A very small quantity dissolved leaving a brown coloured solution. (This may be PbO_2 in suspension.)
(e) Acidified potassium iodide	No reaction.	Brown solution formed.

Experiment 66. Questions

1. $2PbO_2(s) \rightarrow 2PbO(s) + O_2(g)$.

2. (a) PbO_2 is oxidizing chloride ions to chlorine.

$$PbO_2(s) + 4HCl(aq) \rightarrow PbCl_2(s) + 2H_2O(l) + Cl_2(g)$$

  (b) PbO_2 is oxidizing iodide ions to iodine, which dissolves in excess iodide ions.

$$I_2(s) + I^-(aq) \rightleftharpoons I_3^-(aq)$$

3. The yellow liquid contains the complex ion $[PbCl_6]^{2-}$ (hexachloro-plumbate(IV) ion).

4. (a) $SnO_2(s) + 2OH^-(aq) + 2H_2O(l) \rightarrow [Sn(OH)_6]^{2-}(aq)$

  (b) $PbO_2(s) + 2OH^-(aq) + 2H_2O(l) \rightarrow [Pb(OH)_6]^{2-}(aq)$

5. Both oxides (SnO_2 and PbO_2) are amphoteric since they react with acids and alkalis.

6. Oxygen will be evolved when red lead is heated:

$$2Pb_3O_4(s) \rightarrow 6PbO(s) + O_2(g)$$

This reaction can be regarded as the decomposition of the lead(IV) oxide component of red lead i.e. $2PbO_2(s) \rightarrow 2PbO(s) + O_2(g)$

(Continued on next page.)

(Continued from last page.)

7. Tin(II) oxide might be expected to combine with oxygen on heating, to form tin(IV) oxide because the +4 oxidation state is generally more stable for tin. This happens in practice:

$$2SnO(s) + O_2(g) \rightarrow 2SnO_2(s)$$

A similar oxidation of lead(II) oxide to lead(IV) oxide would not be expected because the latter decomposes on heating (see Question 1) and the +2 oxidation is generally more stable for lead.

In practice, prolonged heating in air at about 400 °C causes partial oxidation to dilead(II) lead(IV) oxide, Pb_3O_4, which decomposes again at higher temperatures.

$$6PbO(s) + O_2(g) \rightleftharpoons Pb_3O_4(s)$$

Experiment 67. Specimen results

Results Table 67a. Tests with Q.

Method	Observation	Inference
(1) A spatula-ful of the powder was heated in an ignition tube, gently at first and then more strongly.	The red-orange solid initially darkened on heating and a colourless gas was evolved as the solid melted to a red liquid. On cooling the liquid solidified to a yellow solid.	The red-orange colour of Q suggests that it could be Pb_3O_4 and that the resulting yellow solid is PbO.
A glowing splint was lowered into the tube.	The splint relit as the solid melted.	O_2 produced, probably by the decomposition of Pb_3O_4 to leave PbO.
(2) About 2 cm^3 of dilute nitric acid was added to half a spatula-ful of the powder in a test-tube. The mixture was gently heated.	The orange solid turned to a dark brown solid on heating. The resulting brown solid settled to the bottom on standing with a colourless liquid above it.	The PbO part of Pb_3O_4 ($2PbO \cdot PbO_2$) dissolves leaving brown PbO_2.
(3) The colourless liquid from test 2 was decanted off carefully into a test-tube.		The colourless liquid is likely to contain Pb^{2+}(aq).
(a) About 1 cm^3 was poured into another test-tube and a few drops of KI solution were added.	A yellow precipitate was formed.	Yellow precipitate of PbI_2.
(b) NaOH(aq) was added to half the remaining solution from (2), drop by drop initially and then to excess.	White precipitate formed initially but this redissolved on addition of excess NaOH(aq).	White precipitate of $Pb(OH)_2$ formed which dissolved to form plumbate(II) ions.

Results Table 67b. Experiment to test inference.

Inference	Test and observation	Conclusion
The resulting solution from test (2) contains Pb^{2+}(aq).	A yellow precipitate appeared on adding a few drops of $CrO_4{}^{2-}$(aq) to the unknown solution. The precipitate turned orange on heating.	Yellow $PbCrO_4$ is formed confirming the presence of Pb^{2+}.

Results Table 67c

Test	Observations	Inferences
1. To 1 cm³ of the solution of H add aqueous silver nitrate followed by dilute nitric acid.	White precipitate appeared on adding silver nitrate with no further change on adding nitric acid.	AgCl(s) probably precipitated. The solution of H is likely to contain Cl^-(aq) ions.
2. To 1 cm³ of the solution of H add aqueous sodium hydroxide until in excess.	White precipitate appeared which dissolved on addition of excess NaOH.	The precipitated hydroxide is amphoteric. Thus, any of the following metal ions could be present: Pb^{2+}, Zn^{2+}, Al^{3+}, Sb^{3+}, Sn^{2+} and Sn^{4+}.
3. To 1 cm³ of aqueous iron(III) chloride add a few drops of aqueous potassium thiocyanate. To this solution add some of the solution of H.	A red solution appeared on adding potassium thiocyanate to iron (III) chloride. The solution decolorized on addition of H.	This indicates that the Fe^{3+} has been reduced to Fe^{2+} by the solution of H. Therefore H is likely to contain Sn^{2+} since this is readily oxidized to Sn^{4+}.
4. To 1 cm³ of aqueous mercury(II) chloride add a little of the solution of H, then excess.	A white precipitate initially formed which darkened to a grey solid on standing.	The white precipitate of Hg_2Cl_2 is reduced to mercury by the Sn^{2+}.
5. Heat some of I in a pyrex boiling-tube. Allow to cool. Add 8-10 cm³ of dilute nitric acid to the residue and boil the mixture for 1 or 2 minutes. Filter if necessary and use portions of the cool solution for the following tests:	Dark brown solid melted on heating and gave off a colourless gas which relit a glowing splint. On further heating the liquid became red and cooled to a glassy yellow-brown solid.	Oxygen evolved probably from the decomposition of brown PbO_2 to yellow PbO.
(a) To 1 cm³ of the solution add aqueous sodium hydroxide.	A white precipitate was formed which dissolved on addition of excess NaOH.	Precipitated hydroxide is amphoteric. $Pb(OH)_2$ is probably precipitated which dissolves to give a plumbate(II).
(b) To 1 cm³ of the solution add dilute sulphuric acid.	A white precipitate is formed.	$PbSO_4$ probably precipitated.
(c) To 1 cm³ of the solution add aqueous potassium chromate(VI).	Yellow precipitate formed.	$PbCrO_4$ precipitated. Confirms the presence of Pb^{2+}.

In the reactions 2 to 4 lead is in the +4 oxidation state. In reactions 3 and 4 it is reduced to its more stable oxidation state of +2.

Experiment 68. Specimen results

Table 68a

Ion (hydrated)	VO_2^+	VO^{2+}	V^{3+}	V^{2+}
Colour	Yellow	Blue	Green	Violet
Oxidation state	5	4	3	2
Name	Dioxovanadium(V)	Oxovanadium(IV)	Vanadium(III)	Vanadium(II)

Results Table 68b

Test	Observations	Summary of reaction(s)
Ammonium vanadate + acid	The white solid turned red and dissolved to a yellow solution.	$\overset{+5}{V}O_3^- \rightarrow \overset{+5}{V}O_2^+$
Vanadium(V) + zinc	The zinc effervesced and the yellow solution became green, blue, green again and, eventually, violet.	$\overset{+5}{V}O_2^+ \rightarrow \overset{+4}{V}O^{2+} \rightarrow V^{3+} \rightarrow V^{2+}$
Vanadium(II) + manganate(VII)	The violet solution became green, blue, green, yellow and, finally, pink.	$V^{2+} \rightarrow V^{3+} \rightarrow VO^{2+} \rightarrow VO_2^+$
Vanadium(V) + sulphite. Add vanadium(II)	The yellow solution became blue. On adding V(II) the mixture became green.	$\overset{+5}{V}O_2^+ \rightarrow \overset{+4}{V}O^{2+}$ $\overset{+4}{V}O^{2+} + V^{2+} \rightarrow V^{3+}$
Vanadium(V) + iodide + thiosulphate	The yellow solution became a muddy brown. Addition of thiosulphate gave a clear blue solution.	$\overset{+5}{V}O_2^+ \rightarrow \overset{+4}{V}O^{2+}$
Vanadium(II) + concentrated sulphuric acid	The violet solution became green.	$V^{2+} \rightarrow V^{3+}$

Experiment 68. Questions

1. The first appearance of a green colour is due to a mixture of yellow vanadium(V) and blue vanadium(IV).

2. The subsequent changes are due to continuing reduction.

    Blue $\rightarrow$ green   :   V(IV) $\rightarrow$ V(III)

    Green $\rightarrow$ violet :   V(III) $\rightarrow$ V(II)   (slow)

    These changes are reversed on adding manganate(VII) except that when oxidation is complete the solution is pink.

(Continued on next page.)

(Continued from last page.)

3. The addition of iodide ions turned the solution a muddy brown. Iodine
    was produced in the reaction.

4. Sodium thiosulphate was added to remove iodine and reveal the blue
    colour of vanadium(IV).

5. Since the zinc half-cell has the most negative electrode potential of
    those shown, it can perform all the reduction stages. But the electrode
    potentials show that iodide will reduce V(V) to V(IV) but not V(IV) to
    V(III):

$$2VO_2^+(aq) + 4H^+(aq) + 2I^-(aq) \rightleftharpoons 2VO^{2+}(aq) + 2H_2O(l) + I_2(aq)$$

$$\Delta E^\ominus = 1.00 \text{ V} - 0.54 \text{ V} = +0.46 \text{ V}$$

A positive value for $\Delta E^\ominus$ shows the reaction is feasible (from left to
right as written).

$$2VO^{2+}(aq) + 4H^+(aq) + 2I^-(aq) \rightleftharpoons 2V^{3+}(aq) + 2H_2O(l) + I_2(aq)$$

$$\Delta E^\ominus = 0.34 \text{ V} - 0.54 \text{ V} = -0.20 \text{ V}$$

A negative value for $\Delta E^\ominus$ shows that this reaction is not feasible under
standard conditions.

6. The addition of sulphite ions (which combine with acid to give H_2SO_3)*
    readily reduces V(V) to V(IV) giving a blue solution, but does not cause
    further reduction to V(III) despite the prediction from electrode poten-
    tials:

$$2VO^{2+}(aq) + 4H^+(aq) + H_2SO_3(aq) + H_2O(l) \rightleftharpoons$$

$$2V^{3+}(aq) + 2H_2O(l) + SO_4^{2-}(aq) + 4H^+(aq)$$

which, on cancelling $H^+(aq)$ and $H_2O(l)$, gives:

$$2VO^{2+}(aq) + H_2SO_3(aq) \rightleftharpoons 2V^{3+}(aq) + H_2O(l) + SO_4^{2-}(aq)$$

$$\Delta E^\ominus = 0.34 \text{ V} - 0.17 \text{ V} = +0.17 \text{ V}$$

Note that a greenish colour may appear if excess sulphite is added, but
this is due to the alternative oxidation product $S_2O_6^{2-}$ which decomposes
in acid solution to give sulphur. Filtering the greenish mixture gives
a blue solution.

7. To oxidize V(II) to V(III) but not V(III) to V(IV) we need a mild oxi-
    dizing agent with an electrode potential between -0.26 V and +0.34 V.
    Sulphuric acid (appearing in the table as SO_4^{2-} + $4H^+$) with $E^\ominus = 0.17$ V
    should be suitable, and you can confirm that the careful addition of a
    few drops of concentrated sulphuric acid to a solution of V(II) does
    produce green V(III).

$$2V^{2+}(aq) + SO_4^{2-}(aq) + 4H^+(aq) \rightleftharpoons 2V^{3+}(aq) + H_2SO_3(aq) + H_2O(l)$$

$$\Delta E^\ominus = 0.17 \text{ V} - (-0.26 \text{ V}) = +0.43 \text{ V}$$

This helps to explain why the final stage in the reduction with zinc is
rather slow. A fairly high concentration of sulphuric acid is needed in
the first place to dissolve the ammonium trioxovanadate(V), but the final
reduction of V(III) to V(II) will not occur until the zinc has lowered
the concentration of acid somewhat by forming hydrogen.

*Equations may show $H_2SO_3(aq)$, $2H^+(aq) + SO_3^{2-}(aq)$, or $SO_2(aq) + H_2O(l)$.

Experiment 69. Questions

1. Mn(VI) forms only in alkaline solution. (You may have noticed that the filter paper appears to act as a catalyst for this reaction.)

2. The addition of acid caused disproportionation of Mn(VI) to Mn(VII) (pink solution) and Mn(IV) (brown solid). The reduction in $[OH^-(aq)]$ caused the equilibrium to shift back to the left:

$$2MnO_4^-(aq) + MnO_2(s) + 4OH^-(aq) \rightleftharpoons 3MnO_4^{2-}(aq) + 2H_2O(l)$$

3. Dilution of the Mn(III) solution caused disproportionation by shifting the equilibrium back to the left:

$$4Mn^{2+}(aq) + MnO_4^-(aq) + 8H^+(aq) \rightleftharpoons 4H_2O(l) + 5Mn^{3+}(aq)$$

4. There is no red colouration.

5. The reactants are solids, which often react slowly.

6. The off-white precipitate of manganese(II) hydroxide darkens on standing, especially where it is in contact with the air. This fact suggests that manganese(II) hydroxide is readily oxidized to brown manganese(IV) oxide.

Experiment 69.

Exercise 1

A cell reaction converting Mn(VII) and Mn(IV) to Mn(VI) is obtained by reversing the second half-equation and adding to the first:

$$2MnO_4^-(aq) + MnO_2(s) + 2H_2O(l) \rightleftharpoons 3MnO_4^{2-}(aq) + 4H^+(aq);$$

$$\Delta E^\ominus = -2.26 \text{ V} + 0.56 \text{ V} = -1.70 \text{ V}$$

Since $\Delta E^\ominus$ is large and negative, the reaction will not proceed from left to right to any measurable extent. Indeed, any Mn(VI) obtained by other means would be expected to disproportionate completely in acid conditions. (The same conclusion may be reached by applying the 'anticlockwise rule' to the two half-equations.)

We also see that reducing $[H^+(aq)]$ would shift equilibrium to the right and make the desired reaction more likely. We should therefore consider alkaline conditions.

A cell reaction converting Mn(VII) and Mn(IV) to Mn(VI) is obtained by reversing the third half-equation and adding to the first:

$$2MnO_4^-(aq) + MnO_2(s) + 2OH^-(aq) \rightleftharpoons 3MnO_4^{2-}(aq) + 2H_2O(l); \quad \Delta E^\ominus = -0.04 \text{ V}$$

Here we see that $\Delta E^\ominus$ is negative again, but this time very small, so that we might expect the reaction to produce an equilibrium mixture containing significant proportions of both reactants and products. Furthermore, we should be able to shift the equilibrium to the right by increasing $[OH^-(aq)]$, thus producing more Mn(VI) in the equilibrium mixture.

(Continued on next page.)

(Continued from last page.)

Exercise 2

For acid conditions, combining the first two equations gives:

$$MnO_2(s) + 4H^+(aq) + Mn^{2+}(aq) \rightleftharpoons 2Mn^{3+}(aq) + 2H_2O(l); \quad \Delta E^{\ominus} = -0.56 \text{ V}$$

Since $\Delta E^{\ominus}$ is negative and fairly large, the reaction will not proceed from left to right to any measurable extent under standard conditions. Increasing $[H^+(aq)]$ still further would shift the equilibrium somewhat to the right, but not sufficiently to give an appreciable amount of $Mn^{3+}(aq)$, because $\Delta E^{\ominus}$ is outside the range of reversibility (i.e. -0.4 V to +0.4 V). For alkaline conditions, combining the last two equations gives:

$$MnO_2(s) + 2H_2O(l) + Mn(OH)_2(s) + \cancel{OH^-(aq)} \rightleftharpoons 2Mn(OH)_3(s) + \cancel{OH^-(aq)};$$
$$\Delta E^{\ominus} = +0.30 \text{ V}$$

Since $\Delta E^{\ominus}$ is positive, but less than 0.40 V, we should expect this reaction to proceed from left to right to an equilibrium mixture containing significant amounts of both reactants and products. However, since two of the reactants are solids, we might expect the reaction to be slow.

Since $OH^-(aq)$ does not appear in the final equation, changing its concentration will not change $\Delta E^{\ominus}$.

Exercise 3

Combining the two half-equations (multiplying the first by 5 to equalise the number of electrons) gives:

$$5Mn^{2+}(aq) + MnO_4^-(aq) + 8H^+(aq) \rightleftharpoons Mn^{2+}(aq) + 4H_2O(l) + 5Mn^{3+}(aq);$$

which reduces to: $\quad 4Mn^{2+}(aq) + MnO_4^-(aq) + 8H^+(aq) \rightleftharpoons 4H_2O(l) + 5Mn^{3+}(aq)$

$$\Delta E^{\ominus} = 1.50 \text{ V} - 1.51 \text{ V} = -0.01 \text{ V}$$

$\Delta E^{\ominus}$ is negative, which means that the reaction from left to right is not favoured under standard conditions, but it should be possible to make $\Delta E^{\ominus}$ appreciably positive by increasing the hydrogen ion concentration. This will shift the equilibrium to the right.

Experiment 70. Specimen results

Results Table 70

Ligand	H_2O	Cl^-	$C_2O_4{}^{2-}$ = ox	$NH_2C_2H_4NH_2$ = en	$C_{10}H_{14}O_8N_2{}^{4-}$ = edta					
Complex ion	$Cu(H_2O)_4{}^{2+}$	$CuCl_4{}^{2-}$	$Cu(ox)_2{}^{2-}$	$Cu(en)_2{}^{2+}$	$Cu(edta)^{2-}$					
Colour	Blue	Yellowish-green	Blue	Violet	Light blue					
Stability constant	–	4.0×10^5 $mol^4\ dm^{-12}$	2.1×10^{10} $mol^2\ dm^{-6}$	–	6.3×10^{18} $mol\ dm^{-3}$					
	Predictions (P) and results (R). ✓ = replacement, x = none									
Test	P	R	P	R	P	R	P	R	P	R
Add H₂O			x	✓	x	x		x	x	x
Add Cl⁻	✓	✓			x	x		x	x	x
Add ox	✓	✓	✓	✓				x	x	x
Add en	✓	✓	✓	✓	✓	✓			x	x
Add edta	✓	✓	✓	✓	✓	✓		✓		

Experiment 70. Questions

1. Yes. Generally, polydentate ligands give more stable complexes than
    monodentate ligands, stability increasing with the number of bonds
    formed. This is only a general rule, however, and it is possible to
    have a complex containing, say, strongly bonded monodentate ligands
    which is more stable than one containing weakly bonded bidentate ligands.

2. Since *en* ligands displace *ox* ligands, but not vice versa, and they are
    both bidentate, $Cu(en)_2{}^{2+}$ must have a substantially greater stability
    constant than $Cu(ox)_2{}^{2-}$. It is not directly comparable with the
    stability constant for $Cu(edta)^{2-}$ which has different units because edta
    is hexadentate.

3. H_2O and Cl^- ligands appear to be readily interchangeable when concen-
    trations are changed.

4. $Cu^{2+}(aq) + 4Cl^-(aq) \rightleftharpoons CuCl_4{}^-(aq)$

$$K_{st} = \frac{[CuCl_4{}^{2-}(aq)]}{[Cu^{2+}(aq)][Cl^-(aq)]^4} = 4.0 \times 10^5\ mol^{-4}\ dm^{-12}$$

$$\therefore\quad \frac{[CuCl_4{}^{2-}(aq)]}{[Cu^{2+}(aq)]} = 4.0 \times 10^5\ mol^{-4}\ dm^{-12} \times [Cl^-(aq)]^4$$

(a) $\dfrac{[CuCl_4{}^{2-}(aq)]}{[Cu^{2+}(aq)]} = 4.0 \times 10^5 \times 5.0^4 = \boxed{2.5 \times 10^8}$

(b) $\dfrac{[CuCl_4{}^{2-}(aq)]}{[Cu^{2+}(aq)]} = 4.0 \times 10^5 \times 0.05^4 = \boxed{2.5}$

These figures show that dilution of the solution will soon reduce
$[Cl^-(aq)]$ sufficiently to convert a significant proportion of $CuCl_4{}^{2-}$
ions to $Cu(H_2O)_4{}^{2+}$ ions and so change the colour.

Experiment 71. Specimen results

Results Table 71

Tube number	1	2	3	4	5	6	7	8
Volume of $(NH_4)_2SO_4$/cm³	15	5	5	5	5	5	5	5
Volume of $CuSO_4$/cm³	0	1.0	1.5	2.0	2.5	3.0	4.0	5.0
Volume of NH_3/cm³	0	9.0	8.5	8.0	7.5	7.0	6.0	5.0
Colorimeter reading	0	0.20	0.29	0.34	0.32	0.28	0.22	0.15

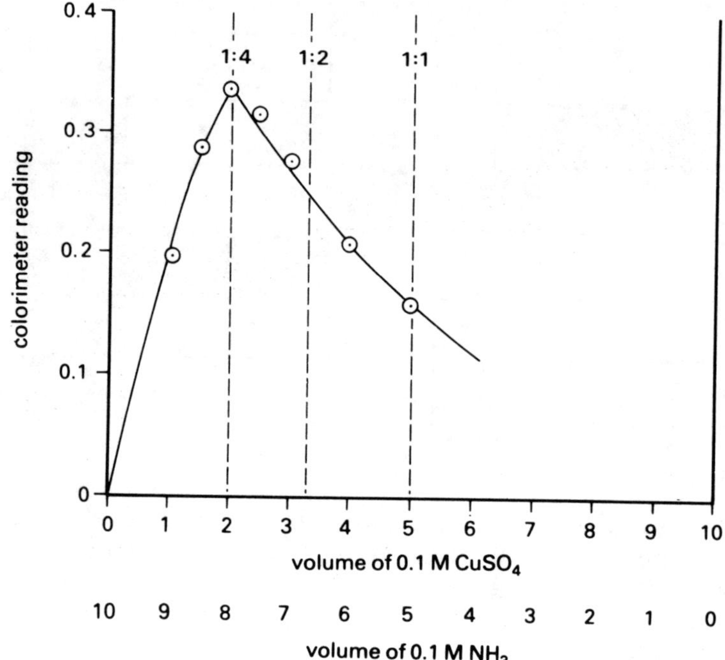

The maximum absorbance for a ratio $CuSO_4$: NH_3 = 1 : 4 indicates that the formula is $Cu(NH_3)_4^{2+}$.

Experiment 71. Questions

1. The addition of ammonia solution to copper sulphate solution usually
    causes precipitation of pale blue copper hydroxide:

    $$NH_3(aq) + H_2O(l) \rightleftharpoons NH_4^+(aq) + OH^-(aq)$$

    or $\qquad$ $Cu^{2+}(aq) + 2OH^-(aq) \rightarrow Cu(OH)_2(s)$

    However, the addition of ammonium sulphate reduces $[OH^-(aq)]$ so much,
    by the common ion effect, that precipitation does not occur.

2. Although the amount of copper present increases, there is less and less
    ammonia present to convert it into the highly coloured complex ion. The
    concentration of the complex ion and, therefore, the absorbance,
    decreases.

3. Although the amount of ammonia decreases, there is always an excess up
    to the peak, so that the concentration of complex ion increases as the
    amount of copper present increases.

4. In the unlikely event of your choosing a filter which was selective
    enough to pass only light which could be absorbed by $Cu(H_2O)_4^{2+}$ and not
    by $Cu(NH_3)_4^{2+}$, absorbance readings would increase throughout the experi-
    ment. In practice, however, both filtering and absorbing occur over a
    band of wavelengths, and a certain amount of absorbance will occur using
    any filter. Using the wrong filter in this experiment would still give
    the same final result although the peak would be less pronounced. More
    care may be necessary in other experiments where species of different
    colours with similar intensities may be present.

5. Since the colour of $Cu(NH_3)_4^{2+}(aq)$ is far more intense than that of
    $Cu(H_2O)_4^{2+}(aq)$, reasonable results can be obtained in this experiment by
    using no filter at all! You were asked to use a filter to establish the
    general principle, which may be far more important in other similar
    experiments.

Experiment 72. Questions – parts A and B

1. The white solid is copper(I) chloride, CuCl. Colour in the transition
    elements is associated with partly filled d-orbitals. Cu(I) has a partly
    filled set of d-orbitals - $1s^2 2s^2 2p^6 3s^2 3p^6 3d^{10}$.

2. In part A, the reducing agent is copper(0), in part B it is sulphur(IV).

3. Heat was required in part A because the reaction involved a solid - such
    reactions are often slow at room temperature.

4. The excess of chloride ions stabilizes the Cu(I) oxidation state in a
    complex ion so that reduction of Cu(II) to Cu(I) can proceed in solution.

5. The dark brown colour is due to the increasing concentration of the
    copper(I) complex $CuCl_4^{3-}$ as it is formed from the green/yellow $CuCl_4^{2-}$.

6. $Cu^+(aq)$, present in very small concentration, is oxidized by air to
    $Cu^{2+}(aq)$. The equilibrium

    $$CuCl(s) \rightleftharpoons Cu^+(aq) + Cl^-(aq)$$

    is then shifted to the right to allow further oxidation of $Cu^+(aq)$ to
    $Cu^{2+}(aq)$, eventually giving a blue solution.

Experiment 72. Questions – part C

1. A muddy brownish suspension is formed. On settling, the precipitate
    appears creamy-coloured and the solution brown.

2. The thiosulphate removes the brown coloration, leaving a clear solution
    over a creamy precipitate. The brown coloration must therefore have
    been due to iodine.

3. Since iodide has been oxidized to iodine, copper(II) must have been
    reduced to copper(I).

4. The creamy-white precipitate must therefore be copper(I) iodide, CuI.

Results Table 72

	Method	Observations	Equation(s)
A. Preparation of copper(I) chloride.	Heat copper(II) chloride, copper and excess sodium chloride in a little water. Dilute.	The green solution darkened slowly to almost black. A white ppt. appeared on dilution.	$CuCl_2(aq) + Cu(s)$ $\rightarrow 2CuCl(s)$ (via $CuCl_4{}^{2-}$ and $CuCl_4{}^{3-}$)
B. Preparation of copper(I) chloride.	Mix solutions of copper(II) chloride and sulphite ions.	A white precipitate appeared.	$2CuCl_2(aq) + SO_3{}^{2-}(aq)$ $+ H_2O(l) \rightarrow 2CuCl(s)$ $+ SO_4{}^{2-}(aq) + 2HCl(aq)$
C. Preparation of copper(I) iodide	Mix solutions of copper(II) sulphate and potassium iodide. Add sodium thiosulphate.	A muddy brown suspension was formed. Adding thiosulphate left a cream ppt. and a clear solution.	$2Cu^{2+}(aq) + 4I^-(aq)$ $\rightarrow 2\,CuI(s) + I_2(aq)$ $I_2(aq) + 2S_2O_3{}^{2-}(aq) \rightarrow$ $2I^-(aq) + S_4O_6{}^{2-}(aq)$

Experiment 73. Questions

1. Carbon dioxide from the redox reaction, and oxygen from the decomposi-
    tion of hydrogen peroxide, which is also catalysed by Co^{2+}.

2. The dark green colour is due to a complex ion of Co(III) with
    2,3-dihydroxybutanedioate ions as ligands. The green colour disappears
    as this complex completes the oxidation of excess 2,3-dihydroxybutane-
    dioate and is itself reduced back to Co(II).

3. The green colour persists for some time at room temperature because the
    reaction between the Co(III) complex and 2,3-dihydroxybutanedioate ion
    is slow.

Experiment 74. Specimen results

Results Table 74

Transition element		Chromium	Manganese	Iron
Oxidation states used		Cr(VI) Cr(III)	Mn(VII) Mn(II)	Fe(III) Fe(II)
Electrode potential/V (higher ox. state $\rightleftharpoons$ lower)		1.33	1.51	0.77
Electrode potential/V ($I_2 + 2e^- \rightleftharpoons 2I^-$)		0.54	0.54	0.54
Electrode potential/V ($S_2O_8{}^{2-} + 2e^- \rightleftharpoons 2SO_4{}^{2-}$)		2.01	2.01	2.01
Does higher ox. state oxidize I^-?	Prediction	Yes	Yes	Yes
	Practically	Yes	Yes	Yes
Does lower state reduce $S_2O_8{}^{2-}$?	Prediction	Yes	Yes	Yes
	Practically	No	No	No
Do you expect catalysis?		No	No	Yes
Time for blue colour to appear/s		Varies with conditions		

Experiment 74. Questions

1. Iron is an effective catalyst, both as Fe(II) and Fe(III). A simplified
    mechanism might consist of two steps, either of which could come first:

    Fe(II) reacts with peroxodisulphate $\rightarrow$ Fe(III) and sulphate.

    Fe(III) reacts with iodide $\rightarrow$ Fe(II) and iodine.

2. Chromium cannot function as a catalyst because Cr(III) will not react
    with peroxodisulphate. This is clearly a kinetic effect, because
    $\Delta E^{\ominus}$ is favourable.

3. Manganese(VII) functions well as a catalyst even though Mn(II) will not
    react with peroxodisulphate. Another oxidation state - Man(III), Mn(IV)
    or Mn(VI) - could be produced as an intermediate.

4. Laboratory temperatures can vary considerably over two days (or even over
    a double period, or in different parts of the room!) Temperature changes
    have a marked effect on rates of reaction and could make it difficult to
    decide whether catalysis has occurred.

Experiment 75. Specimen results

Results Table 75

Test	Observations	Inferences
1. Warm three-quarters of your sample of C with 4-5 cm³ of aqueous sodium hydroxide.	The solid dissolved to a colourless solution (1). A pungent smell was noted on warming (1). The gas turned red litmus blue and gave white smoke in HCl fumes (1).	Ammonia released (1). C is an ammonium salt (1)
2. Add approximately 10 cm³ of dilute sulphuric acid to the remainder of your sample of C in a boiling-tube and warm the mixture in order to dissolve the solid. Use portions of the solution for the following tests:	The white solid turned orange (1) and dissolved to a yellow solution (1).	C may contain a transition metal (1).
(a) To 1-2 cm³ of the solution of C add an equal volume of aqueous potassium iodide. Then add, dropwise, aqueous sodium thiosulphate until there is no further change.	The solution turned dark brown (1). Addition of $Na_2S_2O_3$ gave a blue solution (1).	Iodide oxidized to iodine (1) which was removed by $Na_2S_2O_3$ (1). Colour changes in C suggest the presence of a transition metal (1).
(b) To 2 cm³ of the solution add a little copper powder and warm.	The solution turned from yellow to green (1) and then to blue (1).	C is reduced by copper (1). Possibly vanadium changing from +5 to +4 (1).
(c) To 3-4 cm³ of the solution add a little zinc powder and allow the mixture to stand. It is suggested that you make observations for about 5 minutes and then allow the mixture to stand for a further 30 minutes, making observation from time to time. During this time you should proceed with other tests.	The zinc effervesced (1). The solution slowly turned green (1), then blue (1), then green again (1) and eventually violet (1). The gas popped when ignited (1).	Hydrogen evolved (1). C is reduced through several oxidation states (1). Probably vanadium +5 → +4 (blue) (1) → +3 (green) (1) → +2 (violet)(1)
3. Carry out a flame test on substance D.	The flame turned lilac (1).	D is a potassium salt (1).
4. Dissolve some D in the minimum quantity of distilled water and use portions of the solution for the following tests:	An orange solution was formed (1).	D may contain a transition metal (1).
(a) Add a few drops of this solution to a mixture of 1-2 cm³ of aqueous potassium iodide with an equal volume of dilute sulphuric acid. Then add aqueous sodium thiosulphate dropwise.	The solution turned dark brown (1). Addition of $Na_2S_2O_3$ gave a blue-green solution (1).	Iodine oxidized to iodine (1) which was removed by $Na_2S_2O_3$ (1). Colour changes in D suggest the presence of a transition metal (1).
(b) To 1 cm³ of the solution of D add aqueous sodium hydroxide. Then add dilute sulphuric acid until in excess.	The solution turned yellow (1) and, on addition of acid, became orange (1).	D is a dichromate (1) which changes reversibly to a chromate (1).
(c) To about 2 cm³ of the solution of D add an equal volume of dilute sulphuric acid. Then add hydrogen peroxide solution dropwise until there is no further change.	The first drop gave a deep blue colour (1). This disappeared (1) with effervescence (1) which increased with more H_2O_2 (1). The solution darkened and then became blue-green (1).	(No inference required)
(d) Transfer the solution from (c) to a boiling-tube and add about 10 cm³ of aqueous sodium hydroxide and 1 cm³ of hydrogen peroxide solution. Heat the resulting solution.	A green ppt. (1) was formed which dissolved to a green solution (1). Warming with H_2O_2 gave a yellow soln. (1) and a little bubbling (1).	D is reduced in acid solution by H_2O_2 but oxidized again in alkali (1). Some H_2O_2 decomposed to oxygen (1).

Experiment 75. Questions

D contains the dichromate(VI) ion, $Cr_2O_7{}^{2-}$.

In reaction 4(a) the dichromate(VI) is reduced to Cr^{3+} (oxidation state + III).

In reaction 4(b) the yellow chromate(VI) ion, $CrO_4{}^{2-}$ is formed (oxidation state + VI).

In reaction 4(d) the final yellow solution contains chromium in the + VI oxidation state.

Experiment 76. Specimen results

Results Table 76a Reactions of aluminium

Reagent	Observations	Identity of any gas given off
Sodium hydroxide solution	Slow evolution of gas at first, becoming quicker.	Hydrogen
Sodium hydroxide solution after immersion in $CuCl_2$(aq)	Rapid evolution of gas.	Hydrogen
Dilute hydrochloric acid	No reaction in cold. Gas evolved on heating.	Hydrogen
Dilute hydrochloric acid after immersion in $CuCl_2$(aq)	Gas evolved in the cold.	Hydrogen
Air after immersion in $CuCl_2$(aq)	No visible reaction.	–
Air after immersion in $HgCl_2$(aq)	White crumbly solid appeared on surface.	–

Results Table 76b Reactions of aluminium solutions

	Observations	
Reagent	Solution of aluminium in sodium hydroxide	Solution of aluminium in dilute hydrochloric acid
Dilute sulphuric acid	A white gelatinous precipitate appeared which dissolved on shaking. More acid made the ppt. permanent and still more redissolved it. The tube became warm.	NONE
Sodium hydroxide solution	NONE	A white gelatinous precipitate appeared which dissolved on shaking. More alkali made the ppt. permanent and still more redissolved it. The tube became warm.
Sodium carbonate solution	NONE	A gas was evolved (CO_2) and a white gelatinous precipitate appeared which dissolved on shaking. More alkali made the ppt. permanent but it did not redissolve. The tube became warm.

Experiment 76. Questions

1. Aluminium is normally covered in a thin invisible layer of aluminium
   oxide which is not easily removed and protects the metal surface from
   further attack by air and even by dilute acids. Any oxide formed on
   the surface of iron does not adhere well enough to protect the metal.
   (Remember also that $E^{\ominus}$ values refer only to standard conditions.)

2. Solutions of copper(II) chloride and mercury(II) chloride remove the
   oxide layer and expose the aluminium surface, which then reacts more
   vigorously with acids and alkalis. (How the oxide layer is removed is
   not clear. It is interesting to note that the chloride ions play a part
   - other copper salts are not effective.)

3. Displacement reactions occur with the cations in solution, producing
   metallic copper and mercury. The copper is simply washed off and the
   aluminium surface then becomes again covered with a protective layer of
   oxide. On the other hand, liquid mercury dissolves some aluminium to
   form an alloy (called an amalgam) which exposes aluminium atoms to
   attack by oxygen but does not allow a coherent protective layer to form.

4. Washing-soda and many oven-cleaners are alkaline in solution and therefore tend to attack aluminium pans and other kitchenware.

5. (a) Reaction with dilute acids:

$$2Al(s) + 6H^+(aq) \rightarrow 2Al^{3+}(aq) + 3H_2(g)$$

   (b) Reaction with dilute alkalis:

$$2Al(s) + 2OH^-(aq) + 6H_2O(l) \rightarrow 2Al(OH)_4^-(aq) + 3H_2(g)$$

6. The addition of acid first neutralizes excess alkali, allowing a
   precipitate of aluminium hydroxide to appear. The precipitate dissolves
   in excess acid.

   The addition of sodium hydroxide first neutralizes excess acid, and then
   a precipitate of aluminium hydroxide appears. The precipitate dissolves
   in excess alkali to form aluminate ions.

$$Al^{3+}(aq) + 3OH^-(aq) \rightarrow Al(OH)_3(s)$$

$$Al(OH)_3(s) + OH^-(aq) \rightleftharpoons Al(OH)_4^-(aq)$$

   The addition of alkali first neutralizes excess acid, and then a
   precipitate of aluminium hydroxide appears. The precipitate dissolves
   in excess alkali.

   The addition of sodium carbonate first neutralizes excess acid,
   releasing carbon dioxide. Sodium carbonate is alkaline by hydrolysis:

$$CO_3^{2-}(aq) + H_2O(l) \rightleftharpoons HCO_3^-(aq) + OH^-(aq)$$

   so that a precipitate of aluminium hydroxide again appears. However,
   the concentration of hydroxide ions is never sufficient to redissolve
   the precipitate.

Experiment 77. Questions

1. Anodized aluminium is (a) less shiny than non-anodized aluminium,
    (b) a poor conductor on the surface, and (c) easily dyed.

2. The anodized layer adds some resistance to the circuit. The resistance
    of the solution increases as the temperature falls because the ions move
    less readily.

3. (a) Anodized aluminium is used to increase resistance to corrosion, e.g.
         in window frames, and for articles made more attractive to the
         consumer by dyeing, e.g. saucepan lids.

    (b) Anodized aluminium is not used for articles where electrical
         contacts need to be made.

Experiment 78. Specimen results

Results Table 78a Reactions of nitric acid and nitrates

Reactants	Observations	Identity of gas
1. Copper and 16 M HNO_3	Rapid evolution of red-brown gas. Tube became hot. Green solution formed.	Nitrogen dioxide, NO_2
2. Copper and 8 M HNO_3	Slow bubbling. Gas was colourless near solution, red-brown near mouth of tube. Solution turned blue.	Nitrogen monoxide, NO. Then $NO \rightarrow NO_2$
3. Magnesium and 2 M HNO_3	Rapid bubbling. Gas was brownish but also popped with a lighted splint.	Nitrogen dioxide and hydrogen, NO_2 and H_2
4. Magnesium and 0.5 M HNO_3	Slow bubbling. Gas was colourless and popped with a lighted splint.	Hydrogen, H_2
5. Devarda's alloy and a nitrate in alkali.	Rapid bubbling. Gas was colourless, popped with a lighted splint and turned red litmus paper blue.	Hydrogen and ammonia, H_2 and NH_3.
6. Aluminium foil and a nitrate in alkali.	Slow bubbling at first, then faster. Gas was colourless, flammable and turned red litmus paper blue.	Hydrogen and ammonia, H_2 and NH_3.

Note that the reaction of nitric acid with a metal always gives a mixture of gaseous products, e.g. NH_3, N_2, N_2O, NO, NO_2, H_2 (and N_2H_4, NH_4^+). One product may be predominant under particular conditions. Mg appears to be the only metal to give H_2 but the 'pop' test may not always work in the presence of other gases.

Results Table 78b Reactions of nitrous acid and nitrites

Reactants in solution	Observations	Identity of coloured product(s)	Oxidation or reduction of NO_2^-
7. Sodium nitrite and sulphuric acid	Solution turned blue. Slow effervescence to give brownish gas.	HNO_2 or perhaps N_2O_3 Gas = NO_2	None Oxidation
8. Warm nitrous acid	Blue colour disappeared. Rapid bubbling. Gas was colourless near solution, brown near mouth of tube.	NO_2 (from $NO + O_2$)	Oxidation (and reduction – see Q.4)
9. Potassium iodide and nitrous acid	Brown solution formed and black precipitate. Trace of brownish gas.	Iodine (I_2 & I_3^-) NO_2	Reduction
10. Potassium manganate(VII) & nitrous acid	Solution became colourless. (Trace of brown ppt. which then cleared.)	(MnO_2 due to local deficiency of acid.)	Oxidation
11a. Iron(II) sulphate & sodium hydroxide	Green precipitate formed.	$Fe(OH)_2$	None
11b. Iron(II) sulphate, nitrous acid & sodium hydroxide	Mixture turned black, then gave a yellow solution. Addition of alkali gave a red-brown ppt.	$Fe(H_2O)_5(NO)^{2+}$ Fe^{3+} $Fe(OH)_3$	None Reduction None
12. Aluminium and a nitrite in alkali	Rapid bubbling. Gas was colourless, popped with a lighted splint and turned red litmus blue.	None	Reduction

Experiment 78. Questions

1. Concentrated nitric acid:

$$Cu(s) + 4HNO_3(aq) \rightarrow Cu^{2+}(aq) + 2NO_3^{2-}(aq) + 2NO_2(g) + 2H_2O(l)$$

The red-brown gas is nitrogen dioxide, NO_2. The blue-green solution is concentrated aqueous copper nitrate.

Moderately concentrated nitric acid:

$$3Cu(s) + 8HNO_3(aq) \rightarrow 3Cu^{2+}(aq) + 6NO_3^{2-}(aq) + 2NO(g) + 4H_2O(l)$$

The blue solution is dilute aqueous copper nitrate. The colourless gas is nitrogen monoxide, NO, which reacts with oxygen in the air to produce nitrogen dioxide, NO_2.

$$2NO(g) + O_2(g) \rightarrow 2NO_2(g)$$

(Continued on next page.)

2. The reaction between magnesium and very dilute nitric acid is a simple electron transfer beteen metal and hydrogen, as occurs with other dilute acids.

$$Mg(s) + 2H^+(aq) \rightarrow Mg^{2+}(aq) + H_2(g)$$

At higher concentrations of acid, the combination of nitrate and hydrogen ions becomes a more powerful oxidant than hydrogen ions alone and is reduced by the magnesium to give water and oxides of nitrogen.

3. The main reason is probably that the finely divided alloy presents a greater surface area for reaction than the aluminium foil. Another reason may be that the alloy is not protected initially by a coherent film of aluminium oxide.

4. $$3HNO_2(aq) \rightarrow H^+(aq) + NO_3^-(aq) + 2NO(g) + H_2O(l)$$

This reaction is an example of disproportionation. The oxidation state of nitrogen both increases and decreases, i.e. it changes from +3 to +5 and +2. Nitric acid cannot undergo disproportionation because the oxidation state of nitrogen is already at its maximum value of +5.

5. (a) Both nitrates and nitrites in alkaline solution react with aluminium or Devarda's alloy to produce ammonia. (Ammonium salts, of course, react similarly but can easily be recognised because ammonia is produced by the action of alkali alone.)

   (b) Nitrates do not react with dilute sulphuric acid. Nitrites do react to produce nitrogen monoxide, which immediately combines with oxygen to give red-brown fumes of nitrogen dioxide.

Experiment 79. Specimen results

Results Table 79

Test	Na_2SO_3	Na_2SO_4	$Na_2S_2O_3$	$Na_2S_2O_8$	'X'
1. Warm with dilute hydrochloric acid.	No reaction in cold. Bubbles of gas on warming. Choking smell. $K_2Cr_2O_7$ paper turned green. (SO_2 produced.)	No visible reaction.	Solution turned slightly cloudy, denser when warm. Choking smell. $K_2Cr_2O_7$ paper turned green. (SO_2 produced.)	Gas evolved on warming - has characteristic smell of chlorine and bleaches litmus.	Solution turned slightly cloudy, denser when warm. Choking smell. $K_2Cr_2O_7$ paper turned green. (SO_2 produced.)
2. Add silver nitrate solution.	Initial white ppt. dissolved on shaking. With more $AgNO_3$ a dense white ppt. remained.	No reaction at first, then a faint white ppt. appeared.	Initial white ppt. dissolved on shaking. With more $AgNO_3$ the ppt. remained and turned yellow, brown and black.	Blackish ppt. formed slowly.	Initial white ppt. dissolved on shaking. With more $AgNO_3$ the ppt. remained and turned yellow, brown and black.
3. Add iodine solution (in aqueous potassium iodide).	The brown colour was immediately discharged. (Iodine reduced.)	No visible reaction.	The brown colour was immediately discharged. (Iodine reduced.)	The brown colour became darker. (Iodide oxidized.)	The brown colour was immediately discharged. (Iodine reduced.)
4. Add potassium iodide solution.	No visible reaction.	No visible reaction.	No visible reaction.	A dark brown solution was formed. (Iodide oxidized.)	No visible reaction.
5. Add iron(III) chloride solution and dilute acid. Warm and add sodium hydroxide solution.	A dark red-brown solution was formed which became almost colourless when hot. Addition of alkali gave a green ppt. (Fe^{3+} reduced.)	A yellow solution was formed which darkened a little on warming. Addition of alkali gave a red-brown ppt. (Fe^{3+} not reduced.)	A dark purple solution was formed which cleared when hot and then became cloudy. Addition of alkali gave a green ppt. (Fe^{3+} reduced.)	A yellow solution was formed which darkened a little on warming. Addition of alkali gave a red-brown ppt. (Fe^{3+} not reduced.)	A dark red-brown solution was formed which became almost colourless when hot. Addition of alkali gave a green ppt. (Fe^{3+} reduced.)
6. Heat a small portion of the solid salt.	Crystals turned white & gave off a steamy vapour which condensed on the upper tube (H_2O). White residue turned yellow on strong heating.	A colourless liquid was rapidly formed, which boiled to give a steamy vapour (H_2O) and a white residue. No further reaction.	A colourless liquid was rapidly formed which boiled to give a steamy vapour (H_2O). The yellowish residue turned brown & gave a black viscous liquid.	The solid melted to a colourless liquid. Bubbles of gas relit a glowing splint. (O_2.)	

1. (a) Sulphur dioxide is released from 'sulphurous acid' on heating.

$$\overset{+4}{S}O_3{}^{2-}(aq) + 2H^+(aq) \rightarrow \overset{+4}{S}O_2(g) + H_2O(l)$$

(b) No reaction.

(c) Sulphur is slowly precipitated and sulphur dioxide released.

$$\overset{+2}{S_2}O_3{}^{2-}(aq) + 2H^+(aq) \rightarrow \overset{+4}{S}O_2(g) + \overset{0}{S}(s) + H_2O(l)$$

(d) No reaction.

(e) Silver sulphite is precipitated, but this dissolves initially in excess sulphite to form a complex ion.

(i) $\overset{+4}{S}O_3{}^{2-}(aq) + 2Ag^+(aq) \rightarrow Ag_2\overset{+4}{S}O_3(s)$

(ii) $Ag_2\overset{+4}{S}O_3(s) + \overset{+4}{S}O_3{}^{2-}(aq) \rightarrow 2[Ag\overset{+4}{S}O_3]^-$

(f) Silver sulphate is precipitated, but this is moderately soluble and so the precipitate is not heavy.

$$Ag^+(aq) + SO_4{}^{2-}(aq) \rightarrow Ag_2SO_4(s)$$

(g) Silver thiosulphate is precipitated but this dissolves initially in excess thiosulphate to form a complex ion. The precipitate decomposes to black silver sulphide.

(i) $\overset{+2}{S_2}O_3{}^{2-}(aq) + 2Ag^+(aq) \rightarrow Ag_2\overset{+2}{S_2}O_3(s)$

(ii) $Ag_2\overset{+2}{S_2}O_3(s) + 3\overset{+2}{S_2}O_3{}^{2-}(aq) \rightarrow 2[Ag(\overset{+2}{S_2}O_3)_2]^{3-}$

(iii) $Ag_2\overset{+2}{S_2}O_3(s) + H_2O(l) \rightarrow Ag_2\overset{-2}{S}(s) + 2H^+(aq) + \overset{+6}{S}O_4{}^{2-}(aq)$

(h) It is not clear what happens in this reaction. It is probably similar to (g) with an unstable complex ion producing silver sulphide.

(i) Iodine is reduced to iodide ions and sulphite ions are oxidized to sulphate ions

$$I_2(aq) + \overset{+4}{S}O_3{}^{2-} + H_2O(l) \rightarrow 2I^-(aq) + \overset{+6}{S}O_4{}^{2-}(aq) + 2H^+(aq)$$

(j) No reaction.

(k) Iodine is reduced to iodide ions and sulphite ions are oxidized to tetrathionate ions.

$$I_2(aq) + 2\overset{+2}{S_2}O_3{}^{2-}(aq) \rightarrow 2I^-(aq) + \overset{+2.5}{S_4}O_6{}^{2-}(aq)$$

(As you should recall, the overall oxidation number of 2.5 for S in $S_4O_6{}^{2-}$ arises because two sulphur atoms have oxidation number -1 and two have $+6$.)

(l) Iodide ions (from the potassium iodide) are oxidized to iodine and peroxodisulphate ions are reduced to sulphate ions.

$$2I^-(aq) + \overset{+6}{S_2}O_8{}^{2-}(aq) \rightarrow I_2(aq) + 2\overset{+6}{S}O_4{}^{2-}(aq)$$

[Sulphur in $S_2O_8{}^{2-}$ <u>appears</u> to have an oxidation number of $+7$, given by $\frac{1}{2}(-2 + 16) = 7$. However, as the name suggests, the ion contains an O—O link, which means that two of the oxygen atoms have an oxidation number of -1 and it is these atoms which are reduced.]

(m), (n) and (o) No reaction.

Experiment 79. Questions

(Continued from last page.)

(p) The same reaction as in (1).

(q) An unstable complex ion of uncertain composition is formed initially
     but this decomposes. Iron(III) ions are reduced to iron(II) ions and
     sulphite ions are oxidized to sulphate ions. The green precipitate is
     iron(II) hydroxide.

$$2Fe^{3+}(aq) + \overset{+4}{S}O_3^{2-}(aq) + H_2O(l) \rightarrow 2Fe^{2+}(aq) + \overset{+6}{S}O_4^{2-}(aq) + 2H^+(aq)$$

$$Fe^{2+}(aq) + 2OH^-(aq) \rightarrow Fe(OH)_2(s)$$

(r) and (t) No reaction initially. The red-brown precipitate is iron(III)
             hydroxide.

$$Fe^{3+}(aq) + 3OH^-(aq) \rightarrow Fe(OH)_3(s)$$

(s) An unstable complex ion of uncertain composition is formed initially.
     Iron(III) ions are reduced to iron(II) ions and thiosulphate ions are
     oxidized to tetrathionate ions. The green precipitate is iron(II)
     hydroxide.

$$2Fe^{3+}(aq) + 2\overset{+2}{S}_2O_3^{2-}(aq) \rightarrow 2Fe^{2+}(aq) + \overset{+2\cdot5}{S}_4O_6^{2-}(aq)$$

$$Fe^{2+}(aq) + 2OH^-(aq) \rightarrow Fe(OH)_2(s)$$

(u) Water of crystallization is released. There is possibly some slight
     decomposition to sulphur (and sulphur dioxide?).

(v) Water of crystallization is released rapidly, in sufficient quantity to
     dissolve the salt. The anhydrous sulphate does not decompose at the
     temperature of a Bunsen burner.

(w) Water of crystallization is released rapidly, sufficient to dissolve
     the salt. The anhydrous thiosulphate decomposes, releasing sulphur,
     which melts to a thick black liquid.

(x) The peroxodisulphate releases oxygen on heating and forms the sulphate
     and sulphur dioxide.

$$\overset{+6}{S}_2O_8^{2-}(aq) \rightarrow \overset{+6}{S}O_4^{2-}(aq) + O_2(g) + \overset{+4}{S}O_2(g)$$

2. Reaction (c), between thiosulphate and hydrogen ions, is dispropor-
    tionation. [Also (g) (iii) and (w).]

3.   (i) The peroxodisulphate ion is the strongest oxidizing agent.

      (ii) The sulphite ion and the thiosulphate ion are strong reducing
            agents. [Thiosulphate is stronger, under standard conditions,
            but this is not evident in these reactions.]

     (iii) The sulphate ion is the most stable - it remains intact through-
            out these tests.

4. Dissolve the two salts in water. Add dilute hydrochloric acid and
    barium chloride solution to each. The sulphate gives a dense white
    precipitate of barium sulphate.

$$Ba^{2+}(aq) + SO_4^{2-}(aq) \rightarrow BaSO_4(s)$$

    Barium sulphite is soluble in dilute acids but there may be a faint
    precipitate due to sulphate impurities formed by oxidation.

5. The reactions of X suggest that thiosulphate ions are formed by heating
    sulphur and sulphite ions.

$$S(s) + SO_3^{2-}(aq) \rightarrow S_2O_3^{2-}(aq)$$

    The reaction takes some time to complete, so the prepared solution still
    contained some sulphite.

Experiment 80. Specimen results

Results Table 80

Test	Observations	Inferences
1. To 2 or 3 cm³ of solution B add a few drops of dilute hydrochloric acid. Now add a few drops of aqueous barium chloride.	A dense white ppt. was formed on adding the barium chloride. (1).	B probably a sulphate (1). Ba^{2+} combines with $SO_4^{2-} \rightarrow BaSO_4(s)$ (1).
2. To 2 or 3 cm³ of solution B add a few drops of aqueous lead ethanoate (lead acetate). Now add an excess of aqueous ammonium ethanoate (ammonium acetate) and shake the mixture.	A dense white ppt. was formed (1). It dissolved in excess ethanoate to a colourless soln. (1).	Ppt. probably $PbSO_4$ (1) which is converted to soluble, covalent, lead (II) ethanoate (1).
3. To about 5 cm³ of solution C in a boiling-tube add about 10 cm³ of dilute hydrochloric acid. Allow the mixture to stand for a minute or two. Cautiously smell the mixture. Warm, if necessary, and test the gas evolved. Describe below how you performed this test. Method: A test-strip was soaked in acidified aqueous dichromate and held at the tube's mouth (2).	A white ppt. developed slowly (1). A pungent smell was noted on warming (1). The paper changed from orange to green (1).	White ppt. probably S (1). Disproportionation (1) of thiosulphate (1) gives S and SO_2 (1). SO_2 confirmed (1). It reduces $Cr_2O_7^{2-}$ to Cr^{3+} (1).
4. (a) In a small beaker place 2 or 3 cm³ of aqueous copper(II) sulphate. Acidify with two drops of dilute sulphuric acid. Now add about 5 cm³ of aqueous potassium iodide. (b) To the mixture obtained in 4(a) add solution C until in excess, swirling well.	A cream ppt. (1) began to settle from a brown solution (1). Cream ppt. remained (1) in colourless soln. (1).	Cu^{2+} oxidizes (1) I^- to I_2 and is reduced to CuI (1). Brown soln. is I_3^- - I_2 in $I^-(aq)$ (1). Thiosulphate reduces I_2 to I^- (1).
5. (a) In a small beaker mix about 2 cm³ of aqueous sodium chloride with an equal volume of aqueous silver nitrate. (b) To the mixture obtained in 5(a) add solution C until in excess, swirling well.	A dense white ppt. was formed (1). The ppt. dissolved (1) to colourless soln. (1).	AgCl ppt. from ionic reaction between $Ag^+(aq)$ and $Cl^-(aq)$ (1). $S_2O_3^{2-}$ and Ag^+ form soluble complex ion (1).
6. Dissolve about 1 g of iron(II) sulphate crystals in dilute sulphuric acid and add a few drops of aqueous potassium thiocyanate. Now add a few drops of solution D.	No reaction at first (1) but addition of D gave a deep blood-red coloration (1).	Fe^{2+} oxidized (1) by D to Fe^{3+} which gives a red complex ion (1) with thiocyanate.
7. (a) In a small beaker place about 5 cm³ of aqueous potassium iodide and a few drops of dilute sulphuric acid. Now add about 5 cm³ of solution D. Warm this mixture. (b) To the mixture obtained in 7(a) add solution C until in excess, swirling well.	A pale yellow solution (1) was formed which darkened to brown on warming (1). Soln. became colourless (1). Faint white ppt. (1).	D oxidizes I^- to I_2 (1). Since D contains S and O it is probably a peroxodisulphate (1). Thiosulphate reduces iodine to iodide (1).
8. Evaporate a small quantity of solution B to dryness and perform a flame test on the residue. Describe below how you do this. Method: A Pt wire was cleaned in conc. HCl & held in a blue flame till no coloration was given. It was then dipped in B moistened with conc. HCl & again held in the flame. (3).	A persistent orange-yellow flame colour was seen (1).	Sodium is present in B and also in C & D. (1). From the conclusions noted above, $B = Na_2SO_4$ (1) $C = Na_2S_2O_3$ (1) $D = Na_2S_2O_8$ (1)

Experiment 81. Specimen results

Results Table 81.   Reactions of alkanes

Reaction	Observations

A.   Combustion
 	
    Appearance of flame.
 	
    Sootiness.
	
	Orange and blue flame.
	
	Very little soot.

B.   Action of bromine
    (in 1,1,1-trichloroethane)
 	
   1.   In dark.
 	
   2.   In light.
 	
    Identification of gas.
	
	1. Liquids mix. No other change.
	
	2. Liquids mix and decolorise. Steamy
	     gas evolved.
	
	Damp blue litmus paper turns red.
	White fumes with ammonia stopper.
	(HBr produced.)

C. Action of bromine water	No reaction. Liquids remain separate. On shaking, the brown colour slowly transfers from the lower aqueous to the upper organic layer.
D. Action of acidified potassium manganate(VII)	No reaction. Liquids remain separate.
E. Action of concentrated sulphuric acid	No reaction. Liquids remain separate.

Experiment 81. Questions

1.   Yes, cyclohexane fails to react with three powerful reagents, and reacts
     with a fourth only in the presence of light.

2.   (a)   Carbon dioxide and water.

    (b) $CH_4(g) + 2O_2(g) \rightarrow CO_2(g) + 2H_2O(l)$; $\Delta H^{\ominus} =$ -890 kJ mol^{-1}

          $2C_2H_6(g) + 7O_2(g) \rightarrow 4CO_2(g) + 6H_2O(l)$; $\Delta H^{\ominus} =$ -3120 kJ mol^{-1}

    or $C_2H_6(g) + 3\tfrac{1}{2}O_2(g) \rightarrow 2CO_2(g) + 3H_2O(l)$; $\Delta H^{\ominus} =$ -1560 kJ mol^{-1}

          $C_3H_8(g) + 5O_2(g) \rightarrow 3CO_2(g) + 4H_2O(l)$; $\Delta H^{\ominus} =$ -2220 kJ mol^{-1}

    (c)   The highly exothermic combustion reactions of alkanes make them
         valuable as fuels.

3.   (a)   In a substitution reaction, one atom (or group of atoms) in a
         molecule is replaced by another atom (or group).

    (b)

$$\text{cyclohexane} + Br_2 \rightarrow \text{bromocyclohexane} + HBr$$

(Continued on next page.)

(Continued from last page.)

    (c) Ultraviolet light. ('Light' alone is NOT good enough. It must be
         ultraviolet light.)

    (d) $CH_4 + Br_2 \rightarrow CH_3Br + HBr$ bromomethane

          $CH_3Br + Br_2 \rightarrow CH_2Br_2 + HBr$ dibromomethane

          $CH_2Br_2 + Br_2 \rightarrow CHBr_3 + HBr$ tribromomethane

          $CHBr_3 + Br_2 \rightarrow CBr_4 + HBr$ tetrabromomethane

Experiment 82. Specimen results

Results Table 82.

Reaction	Observations

A. <u>Combustion</u>
    Appearance of flame.
    Sootiness.
	
	Orange and blue flame.
	Little sootiness.

B. <u>Action of bromine</u>
    (in 1,1,1-trichloroethane)
    1. In dark.
    2. In light.
    Identification of gas
    (if any).
	
	1. Liquids mix and decolorise.
	2. Liquids mix and decolorise.
	No gas is given off.

C. <u>Action of bromine water</u>	Bromine rapidly decolorised. Liquids remain separate.
D. <u>Action of acidified potassium</u> <u>manganate(VII)</u>	Rapid decolorisation. Liquids remain separate.
E. <u>Action of concentrated</u> <u>sulphuric acid</u>	Liquids mix and react vigorously. Colour darkens to charred black mass.

Experiment 82. Questions

1. Yes. A comparison of the results for Experiments 81 and 82 shows more
positive reactions for cyclohexene than cyclohexane, indicating the
greater reactivity of alkenes in general. This is likely to be related
to the fact that less energy is required to break the Π bond in an alkene
than the σ-bond in an alkane.

2. The presence of soot in the flame is an indication of the higher carbon
to hydrogen ratio in alkenes than in alkanes.

3. Unlike alkanes, alkenes decolorise bromine in the absence of light and
acidified potassium manganate(VII) solution.

Experiment 83. Specimen results

Results Table 83

	Time for precipi-tate to appear	Observations
A 1-chlorobutane	8 minutes	Faint still after 30 minutes
B 1-bromobutane	2 minutes	Denser after 5 minutes. Settling beginning after 30 minutes.
C 1-iodobutane	30 seconds	Heavy yellow after 1 minute. Settled after 30 minutes.
D chlorobenzene	-	No precipitate even after 1 hour.

Experiment 83. Questions

1. 1-iodobutane, 1-bromobutane, 1-chlorobutane, chlorobenzene.

2. The trend in bond energy values coincides with the order of the speed of hydrolysis, i.e. $C_4H_9I > C_4H_9Br > C_4H_9Cl > C_6H_5-Cl$. This suggests that the weaker the C-halogen bond, the faster the substitution of the halogen.

3. $CH_3CH_2CH_2CH_2X + H_2O \rightarrow CH_3CH_2CH_2CH_2OH + H^+(aq) + X^-(aq)$ (X = Cl, Br or I)
    butan-1-ol

Experiment 84. Specimen results

Results Table 84

Mass of measuring cylinder + 2-methylpropan-1-ol	50.8 g
Mass of measuring cylinder after emptying	43.7 g
Mass of 2-methylpropan-2-ol	7.1 g
Mass of collecting flask	32.2 g
Mass of collecting flask + 2-chloro-2-methylpropane	37.7 g
Mass of 2-chloro-2-methylpropane	5.5 g

Experiment 84. Questions

1. Amount of $(CH_3)_3COH$ used $= \dfrac{7.1 \text{ g}}{74 \text{ g mol}^{-1}} = 0.096$ mol

    The equation shows 1 mol of $(CH_3)_3CCl$ formed from 1 mol of $(CH_3)_3COH$

    $\therefore$ the maximum possible amount of $(CH_3)_3CCl = 0.096$ mol

    and the maximum mass $= 0.096 \text{ mol} \times 91.5 \text{ g mol}^{-1} = 8.9$ g

2. Percentage yield $= \dfrac{\text{mass obtained}}{\text{theoretical mass}} \times 100 = \dfrac{5.5 \text{ g}}{8.9 \text{ g}} \times 100 = \boxed{62\%}$

3. A stronger alkali would tend to hydrolyse the halogeno-compound back to the alcohol.

Experiment 85. Specimen results

Results Table 85 Reactions of ethanol

Property/Reaction	Observations
A. Solubility of water pH of solution	Completely soluble. pH 6 before and after addition.

B.   Mild oxidation
    (a) Smell
 	
    (b) Fehling's solution
 	
    (c) Silver mirror test
	(a) Pungent smell.
	
	(b) Blue colour changed through green
	    to reddish-brown precipitate.
	(c) A silver mirror developed on the
	    inside of the tube.

C.   Further oxidation
    (a) Smell
    (b) Universal indicator paper
    (c) Sodium carbonate
	(a) Vinegary smell.
	(b) Orange; weak acid, pH 3-4.
	(c) Effervescence (CO_2).

D. Triiodomethane reaction	Pale yellow precipitate. 'Antiseptic' smell.
E. Reaction with sodium	Dissolved quietly; gas evolved popped with lighted splint (hydrogen). Evaporation gave white solid.
F. Esterification	Pleasant, sweet smell.

G.   Dehydration
    (a) Bromine water
    (b) Acidified potassium
        manganate(VII) solution
	(a) Decolorized.
	(b) Decolorized.

Experiment 85. Questions

1.   Reaction B. The mild oxidation of ethanol produced a compound which acts
as a reducing agent by giving positive Fehling's and silver mirror tests.
The reducing agent is ethanal, CH_3CHO.

2.   In these reactions, dichromate(VI) ions are reduced to chromium(III) ions.

$$Cr_2O_7{}^{2-}(aq) + 14H^+(aq) + 6e^- \rightarrow 2Cr^{3+}(aq) + 7H_2O(l)$$

     dichromate(VI)   chromium(III)
     (orange)   (blue-green) (violet if excess
                                                          acid, no $Cr_2O_7{}^{2-}$.)

3.   Reaction C. Further oxidation of ethanol gave a product which turned
universal indicator orange and is therefore acidic. The product smelt
of vinegar - ethanoic acid.

(Continued on next page.)

(Continued from last page.)

4. (a) The use of a higher proportion of ethanol than acidified sodium
       dichromate and distilling off the product as fast as it is formed
       favours the formation of ethanal.

   (b) The use of a higher proportion of acidified sodium dichromate than
       ethanol and refluxing the mixture favours the formation of ethanoic
       acid.

5. The fact that hydrogen gas was detected indicates that CO—H fission must
   have taken place.

6. The fact that the product immediately decolorized bromine water and
   acidified potassium manganate(VII) indicates that a compound with a
   C=C bond (an alkene) was produced.

7. Reaction F produced a pleasant, sweet-smelling compound - an ester.

Experiment 86. Specimen results

Results Table 86

Test	Observations
A. Solubility in water: (a) a little phenol, (b) a lot of phenol, (c) pH of solution.	 Slowly dissolved. No more dissolved. pH = 4.
B. Reaction with sodium hydrogencarbonate.	No reaction.
C. Reaction with sodium hydroxide solution. Subsequent addition of hydrochloric acid.	Phenol dissolved readily - far more soluble than in water. Phenol reappeared in an oily mixture.
D. Reaction with sodium.	Vigorous effervescence. Gas ignited with a 'pop'.
E. Reaction with bromine water.	White precipitate. Brown colour disappeared.
F. Action of neutral iron(III) chloride solution.	Violet coloration.

Experiment 86. Questions

1. The smell of phenol often reminds people of antiseptics and/or disin-
    fectants. It was formerly used in 'carbolic soap', a mild antiseptic,
    and some modern products contain similar compounds, e.g. T.C.P.
    (trichlorophenol).

2. (a) Reaction with neutral iron(III) chloride - phenol gives a violet
         colour, ethanol does not.

    (b) Reaction with bromine water - phenol gives a white precipitate,
         ethanol does not.

    (c) pH of solution - phenol gives an acidic solution in water, ethanol
         does not.

3. Phenol dissociates to a certain extent in water, as indicated by the fact
    that a solution of it in water turns universal indicator orange/red.
    However, with the addition of sodium hydroxide, there is interaction
    between the hydroxide ions and hydronium ions. This results in a shift
    of equilibrium to the right, promoting the formation of soluble phenoxide
    ions.

$$C_6H_5OH + H_2O \rightleftharpoons C_6H_5O^- + H_3O^+$$

4. Test A - pH of solution. Ethanol is not sufficiently acidic to donate
    protons to water and therefore appears neutral in solution.

5. Test B - reaction with sodium hydrogencarbonate. Unlike dilute mineral
    acids, phenol is not sufficiently acidic to liberate carbon dioxide from
    a hydrogencarbonate.

6. Substitution must occur in the benzene ring, and appears to be made
    easier by the presence of the −OH group.

Experiment 87. Specimen results

Results Table 87

Test	Phenylamine	Butylamine	Ammonia
A. (a) Solubility in water.	Apparently insoluble.	Soluble.	Very soluble.
(b) pH of solution.	6-7 (same as water).	11-14	11-14
B. (a) Reaction with hydrochloric acid.	Dense white fumes. Clear solution. Exothermic.	Dense white fumes. Clear solution. Exothermic.	White fumes. Clear solution. Exothermic.
(b) Evaporation of water from product.	Colourless crystals immediately.	Colourless crystals.	Colourless crystals.
C. Reaction with aqueous copper(II) sulphate.	Bright green colour which first darkens, then gives a yellow-green precipitate.	Blue-green precipitate which darkens and then dissolves to a deep blue solution.	Pale blue precipitate which darkens and then dissolves to a deep blue solution.
D. (a) Reaction with nitrous acid below 5 °C.	No apparent reaction apart from slight decomposition of nitrous acid.	Slow bubbling. Gas has no effect on limewater and puts out a lighted splint.	Slow bubbling. Gas has no effect on limewater and puts out a lighted splint.
(b) Reaction of product from (a) with: (i) phenol	Bright yellow precipitate.	Yellowish solution.	No change.
(ii) naphthalen-2-ol	Scarlet red precipitate.	Yellowish solution.	No change.
(c) Effect of heat on product from (a)	Fast bubbling Gas has no effect on limewater and puts out a lighted splint.	Fast bubbling. Gas has no effect on limewater and puts out a lighted splint.	Fast bubbling. Gas has no effect on limewater and puts out a lighted splint.
E. Reaction with bromine water.	Brown colour disappears. White 'woolly' precipitate.	No change.	No change.

1. Butylamine is at least as basic as ammonia, whereas phenylamine is
    weaker.

2. Butylamine is soluble in water due to the polar nature of the N—H bond,
    which leads to hydrogen bonding with water. In phenylamine, the non-
    polar benzene ring diminishes the influence of the polar NH_2 group and
    the position of equilibrium for the dissociation of phenylamine in water
    is well over to the left.

$$C_6H_5NH_2 + H_2O \rightleftharpoons C_6H_5NH_3^+ + OH^-$$

However, phenylamine is very soluble in hydrochloric acid. Since this
is a stronger acid than water, the position of equilibrium is well over
to the right.

$$C_6H_5NH_2 + HCl \rightleftharpoons C_6H_5NH_3^+ + Cl^-$$

3. An arylamine reacts with bromine water to give a white precipitate
    whereas an alkylamine does not.

    An arylamine reacts with cold nitrous acid and an alkaline solution of
    naphthalen-2-ol to give a red dye whereas an alkylamine does not.

Experiment 88. Specimen results

Results Table 88a

Test	Observations	Inferences
1.(a) Place 1 cm³ of C in a test-tube and add an equal volume of water. Now add a little anhydrous sodium carbonate.	Effervescence. Gas turned limewater milky.	CO_2 produced which suggests C is a carboxylic acid.
1.(b) Repeat test 1.(a) using D.	No reaction.	D is not a carboxylic acid.
2.(a) (The reactions 2.(a) and 2.(b) should be performed in a fume cupboard.) Place 2 or 3 cm³ of C in a dry beaker. Add a little phosphorus pentachloride. (CARE!)	Vigorous reaction giving a steamy gas. Gas gave white fumes with ammonia.	HCl produced, which suggests C contains an —OH group.
2.(b) Repeat test 2.(a) using D. (CARE!)	Vigorous reaction giving a steamy gas. Gas gave white fumes with ammonia.	HCl produced, which suggests D contains an —OH group.
3. Mix, in a small beaker, about 2 cm³ of each of C, D and concentrated sulphuric acid. (CARE!) Warm gently but do not boil. Pour the mixture into an excess of aqueous sodium carbonate in a large beaker. Smell the product.	A sweet, fruity smell was noticed.	An ester was probably formed from a carboxylic acid, C, and an alcohol, D.
4. Mix about 5 cm³ of D with an equal volume of aqueous potassium dichromate in a test-tube. Pour the mixture into about 10 cm³ of nearly boiling dilute sulphuric acid in a small beaker. Smell the mixture.	A pungent smell. The solution changed from orange to blue-green.	The dichromate has oxidized D (suspected alcohol) to a pungent-smelling aldehyde.
5. In a small beaker dissolve a few crystals of potassium iodide in about 10 cm³ of D. Add a few drops of aqueous sodium hypochlorite. Warm the mixture but do not boil. Cautiously smell the products.	Yellow precipitate. 'Antiseptic' smell.	Triiodomethane produced, which suggests that D contains the group $CH_3CH(OH)$.
6. Make a solution of E in distilled water and use portions for the following tests: To 2-3 cm³ of the solution add aqueous silver nitrate. Then add dilute nitric acid. Finally add dilute aqueous ammonia.	Bromine decolorized. White precipitate formed. White precipitate. Insoluble in dilute nitric acid but soluble in dilute ammonia.	This suggests that E is either an aromatic amine or a phenol. Precipitate of AgCl suggests E contains Cl^- or a readily hydrolysed —Cl group.
7. Dissolve a little E in about 1 cm³ of concentrated hydrochloric acid and dilute to about 4 cm³ with distilled water. Cool the tube in an ice-water mixture and add a few drops of aqueous sodium nitrite (to be prepared by dissolving sodium nitrite in distilled water). Leave the tube in the ice-water mixture. Dissolve a few crystals of phenol (CARE!) in 7-8 cm³ of aqueous sodium hydroxide, cool this solution, and then add it to the cold solution prepared as above.	A dullish orange precipitate was formed.	An azo-dye was probably formed from a primary aromatic amine. This and the previous tests suggest that E is a salt of a primary aromatic amine, $ArNH_3{}^+Cl^-$.

Results Table 88b

	REASONING
Functional group in D: — OH	Reacts with PCl_5 to evolve HCl, suggesting an — OH group. Since it evolves CO_2 from Na_2CO_3, it must be the — OH group of a carboxylic acid.
Functional group in C: $-C\overset{\displaystyle O}{\underset{\displaystyle OH}{}}$	Reacts with PCl_5 to evolve HCl, suggesting the presence of an — OH group. Since it fails to evolve CO_2 from Na_2CO_3, it must be the — OH group of an alcohol.

The structural features of compound E are as follows. E contains at least one benzene ring, to which is attached an $NH_3^+Cl^-$ group.

Experiment 89. Specimen results

Results Table 89

Test	Observations	
	Ethanal	Propanone
A. Addition reaction with sodium hydrogensulphite*	Tube became hot. Some milkiness which soon disappeared.	Tube became hot. White precipitate formed.
B. Condensation reaction with 2,4-dinitrophenylhydrazine	Orange-yellow precipitate.	Orange-yellow precipitate.
C. Reaction with alkali	Yellow precipitate. Brown oily drops congealed when cool. 'Bad apple' smell.	No change.
D. Oxidation reactions: (a) with acidified dichromate(VI)	Orange solution turned blue/green.	No change.
(b) with Fehling's solution	Blue solution became cloudy green, then formed a reddish precipitate.	No change.
(c) with Tollens' reagent	Grey precipitate.**	No change.
E. Triiodomethane reaction	Pale yellow precipitate.	Pale yellow precipitate.

*Ethanal and methanal, unlike other aldehydes and ketones, do not form white precipitates with sodium hydrogensulphite. Addition products are formed but are so soluble that they rarely crystallize.

**Silver is usually produced too rapidly to form a good silver mirror.

Experiment 89. Questions

1. Tests C and D, i.e. the reaction with alkali, acidified dichromate(VI), Fehling's solution and Tollen's reagent, all serve to distinguish between ethanal and propanone.

2. The reaction between ethanal and sodium hydroxide on warming resulted in a thick resin-like substance which could be a polymer.

3. 2,4-dinitrophenylhydrazine could be used in a general test for carbonyl compounds since both aldehydes and ketones react with it to produce orange precipitates.

Experiment 90. Questions

1. The solvent must dissolve the crude material readily when hot, but only
    to a small extent when cold. Ethanol fits both of these requirements
    for this particular experiment.

2. When the derivative crystallizes, soluble impurities remain in solution
    and are separated from the derivatives by filtration.

3. Some of the derivative is always lost during filtering because the
    filtrate is a saturated solution. Using the minimum amount of solvent
    reduces this loss.

4. Insoluble impurities may be removed by filtering the hot solution before
    crystallization occurs. (The funnel and flask must be heated first, by
    pouring hot solvent through it, in order to prevent crystallization
    occurring in the apparatus.)

5. Isomeric aldehydes and ketones frequently have closely similar boiling-
    points, but the melting-points of their derivatives usually differ
    considerably. (Look, for example, at pentan-2-one and pentan-3-one in
    Table 3.) Furthermore, boiling-points are less readily measured with
    precision than melting-points, and it is generally more difficult to
    obtain liquids in a high state of purity.

Experiment 91. Specimen results

Results Table 91 Reactions of ethanoic acid

Reaction	Observations
A. pH of aqueous solution.	Orange-red – pH 3-4.
B. Reaction with sodium hydrogen-carbonate solution.	Gas evolved which turns limewater milky – CO_2.
C. Reaction with sodium.	Gas evolved which popped with a lighted splint – H_2.
D. Reaction with phosphorus pentachloride.	Steamy gas evolved which gave white fumes with ammonia – HCl.
E. Reaction with 2,4-dinitro-phenylhydrazine.	No change.
F. Triiodomethane reaction.	No change.
G. Action of iron(III) chloride.	Red colour at first – brown ppt. on boiling.

Experiment 91. Questions

1. Ethanoic acid resembles hydroxy-compounds more closely than carbonyl
    compounds. It reacts with sodium and phosphorus pentachloride, typical
    reactions of alcohols, but does not react with 2,4-dinitrophenylhydrazine.

2. Ethanoic acid releases carbon dioxide from sodium hydrogencarbonate, but
    phenol does not. Also, the pH of aqueous ethanoic acid is lower than
    that of aqueous phenol of similar concentration.

3. The presence of the large, non-polar benzene ring outweighs the influence
    of the polar carboxyl group, which participates in hydrogen-bonding with
    water.

Experiment 92. Specimen results

Results Table 92

Test	Observations	Inferences
(a) Dissolve some of I in the minimum quantity of distilled water and use portions of the solution for the following tests.		
(i) To 1-2 cm³ of the solution of I add aqueous sodium hydroxide until in excess.	White precipitate, insoluble in excess reagent.	Insoluble hydroxide precipitated by a Group II cation.
(ii) To 1-2 cm³ of the solution of I add an equal volume of dilute sulphuric acid, shake the tube, and allow the solution to stand.	White precipitate formed slowly which settled slowly to the bottom of the tube.	Insoluble sulphate precipitated by a Group II cation, not Mg^{2+}, Sr^{2+} (soluble). Probably Ca^{2+} since $BaSO_4$ is heavy and very insoluble.
(b) Add a little dilute sulphuric acid to some solid I and heat the mixture.	Pungent smell of vinegar.	Ethanoic acid produced from an ethanoate.
(c) Heat a little solid I in a pyrex test-tube.	Sweetish smell. Solid darkened.	Propanone formed from calcium ethanoate?
(d) Using the apparatus shown, heat a little of I in a test-tube and pass the vapour evolved into 2,4-dinitrophenylhydrazine reagent.	Yellow precipitate formed round the end of the delivery tube.	On heating, I must give a ketone or an aldehyde which forms a condensation product with the reagent.
(e) Heat a little solid J in a pyrex test-tube.	Colourless vapour condensing to drops of liquid at top of tube. Smell of burnt wood? White residue.	J is an organic compound.
(f) To a little solid J add a little concentrated sulphuric acid (CARE!) Warm gently.	Effervescence. Gas had no effect on limewater but, when ignited, burnt with a blue flame.	CO, but not CO_2, shows that J is a methanoate.

I = calcium ethanoate   J = calcium methanoate   (barium acceptable)

Experiment 93.Specimen results

Results Table 93

Reaction	Observations

A.   Reaction with water.
 	
     Product tested with ammonia.
 	
     Product tested with iron(III) chloride.
	Moderate reaction giving steamy fumes.
	
	White fumes, suggesting HCl is produced.
	
	Red coloration, suggesting ethanoic acid is formed.

B.   Reaction with ethanol.
 	
     Product tested with ammonia.
 	
     Smell of product.
	Vigorous reaction with crackling and spitting, giving steamy fumes.
	
	White fumes, suggesting HCl is produced.
	
	Sweetish smell, suggesting an ester is produced.

C. Reaction with ammonia.	Violent reaction, with crackling and spitting, giving a lot of white fumes (NH_4Cl?).
D. Reaction with phenylamine.	Violent reaction with crackling and spitting, giving a lot of white fumes. Some white solid around lower walls of beaker.

Experiment 93. Questions

1.(a) $CH_3COCl + H_2O \rightarrow CH_3CO_2H + HCl$
                          ethanoic acid

  (b) $CH_3COCl + C_2H_5OH \rightarrow CH_3CO_2C_2H_5 + HCl$
                              ethyl ethanoate

  (c) $CH_3COCl + 2NH_3 \rightarrow CH_3CONH_2 + NH_4Cl$
                         ethanamide

  (d) $CH_3COCl + C_6H_5NH_2 \rightarrow CH_3CONHC_6H_5 + HCl$ (HCl reacts with excess amine.)
                             N-phenylethanamide

2.   The order of reactivity of the nucleophiles corresponds with the order of basic strength.

$$NH_3 > C_6H_5NH_2 > C_2H_5OH > H_2O$$

The more basic the nucleophile, the more readily available is the lone pair of electrons to make a bond with ethanoyl chloride.

Experiment 94. Specimen results

Results Table 94

Mass of weighing bottle and phenol	20.73 g
Mass of weighing bottle after emptying	15.82 g
Mass of phenol	4.91 g
Mass of specimen bottle	5.61 g
Mass of specimen bottle and phenylbenzoate	9.71 g
Mass of phenylbenzoate	4.10 g
Melting-point of phenyl benzoate	67-68 °C

Experiment 94. Questions

1. Amount of phenol used $= \dfrac{4.91 \text{ g}}{94 \text{ g mol}^{-1}} = 0.0522$ mol

    The equation shows 1 mol of ester formed from 1 mol of phenol

    $\therefore$ maximum possible mass of ester $= 0.0522$ mol $\times 198$ g mol$^{-1} = 10.3$ g

2. Percentage yield $= \dfrac{\text{mass obtained.}}{\text{theoretical maximum mass}} \times 100 = \dfrac{4.10 \text{ g}}{10.3 \text{ g}} \times 100 = \boxed{40\%}$

3. The sample has a sharp melting-point within 1 °C of the melting-point of
    pure phenyl benzoate. This shows the sample is of high purity.

    If your sample has an imprecise melting-point (i.e. melts over a range
    of 2-3 degrees or more) which is well below 69 °C, there must be
    impurities present.

Experiment 95. Specimen results

Results Table 95

EXPERIMENT	OBSERVATIONS	INFERENCES
(a) To 2 cm³ of the solution of F add 2,4-dinitrophenylhydrazine reagent.	Yellow precipitate formed.	A condensation reaction occurs. This shows F is an aldehyde or ketone.
(b) Prepare a sample of Tollens' reagent as follows: to 5 cm³ of aqueous silver nitrate in a test-tube add 1-2 drops of aqueous sodium hydroxide. Then add dilute aqueous ammonia until only a trace of precipitate remains. Now add 5 drops of the solution of F and place the tube in hot water. (Pour the contents of the tube down the sink on completion of the test.)	No change observed.	No reduction of silver(I) to silver metal. Therefore, F is not an aldehyde.
(c) Add a little of the solution of F to some sodium hydrogencarbonate.	Effervescence. Gas turned limewater cloudy.	Carbon dioxide released by acid. F probably contains a carboxyl group, CO_2H.
(d) To 1 cm³ of the solution of F add 3 cm³ of aqueous potassium iodide and then 10 cm³ of sodium chlorate(I) (sodium hypochlorite) solution.	Yellow precipitate formed.	Triiodomethane produced. This shows the presence of structural group $CH_3C{=}O$ or CH_3CHOH.
(e) Place ONE drop of G on an inverted crucible lid and ignite G from above.	G burned with a sooty luminous flame.	G is probably an aromatic compound.
(f) To 1-2 cm³ of G add 2,4-dinitrophenylhydrazine reagent.	An orange-yellow precipitate was formed.	A condensation reaction occurs. (1) This shows G is an aldehyde or ketone.
(g) Prepare a sample of Tollens' reagent as in (b). Add 5 drops of G, shake the mixture and place the tube in hot water. (N.B. Pour the contents of the tube down the sink on completion of the test.)	A silver mirror was formed slowly on the inner surface of the tube.	Formation of silver metal from silver(I) shows G is a reducing agent — probably an aldehyde.
(h) To 0.5 cm³ of G add 2 cm³ of 6 M sodium hydroxide. Warm the mixture and stir well for 5 minutes. Then add sufficient water to dissolve any residue which has formed. Separate a portion of the aqueous layer and add a little concentrated hydrochloric acid to it. Cool.	A white precipitate was formed on the addition of acid.	Precipitate could be an aromatic acid (benzoic?) formed from an aldehyde by the Cannizzaro reaction. $RCHO \rightarrow RCO_2^- + RCH_2OH$
(i) Test to see if H is miscible with (i) cold water, (ii) hot water.	H appears to be insoluble. Two layers separate on standing. On warming, the two layers merged. Smell of vinegar. Solution turned litmus red.	H is not a lower member of an alcohol, aldehyde, ketone, amine or carboxylic acid series. H could be an anhydride of a carboxylic acid.
(j) (i) To about 2 cm³ of H in a dry beaker add a little anhydrous sodium carbonate. (ii) Now add about 5 cm³ of hot water to the mixture from test (j),(i).	No change observed. Effervescence. Gas made limewater cloudy.	H is not a carboxylic acid. Carbon dioxide released by acid formed by hydrolysis.
(k) (This reaction should be performed in a fume cupboard.) Place 2 or 3 cm³ of H in a dry beaker. Add a little phosphorus pentachloride (CARE!).	Solid dissolved. No gas evolved.	H is not an alcohol or carboxylic acid.
(l) Place 2 cm³ of glacial ethanoic acid and 2 cm³ of H in a boiling tube and add 2 cm³ of phenylamine (aniline). Heat cautiously until the mixture begins to boil. Remove from the flame and allow to stand for 2 minutes. Now pour the mixture into about 125 cm³ of cold water with stirring and allow to stand.	A creamy precipitate was formed.	A substituted amide is formed by a reaction between phenylamine and H, which is probably an anhydride.

Experiment 95. Questions

1. Since F is acidic (test (c)) it probably contains a carboxyl group,
    $-CO_2H$.

    Since F forms a 2,4-dinitrophenylhydrazone (test (a)) it must contain
    a carbonyl group, $>C=O$ in addition to the $-CO_2H$ group.

    Since F undergoes the triiodomethane reaction (test (d)) it must
    contain one of the structural groupings $CH_3\overset{|}{C}=O$ or $CH_3\overset{|}{C}HOH$

2. Test (e) suggests that G is aromatic.

    The formation of a 2,4-dinitrophenylhydrazone in test (f) shows the
    presence of a carbonyl group, $>C=O$, and the formation of a silver
    mirror in test (g) suggests that the carbonyl group is in an aldehyde,
    $-CHO$.

3. H contains the carboxylic anhydride group, $-CO-O-CO-$.

    Test (f) shows that a solution of H in hot water is acidic, while
    test (k) shows that H itself is not a carboxylic acid. This suggests
    that H could be an anhydride, which is confirmed by the formation of
    a substituted amide in test (l).

Experiment 96. Questions

1. The solution would be slightly alkaline. Since $CO_3{}^{2-}$ is the conjugate
    base of the weak acid $HCO_3{}^-$ which is, in turn, the conjugate base of
    the weak acid H_2CO_3, the following equilibria would be set up:

    $$CO_3{}^{2-}(aq) + H^+(aq) \rightleftharpoons HCO_3{}^-(aq)$$

    $$HCO_3{}^-(aq) + H^+(aq) \rightleftharpoons H_2CO_3(aq)$$

    These would decrease the concentration of hydrogen ions in solution,
    making it slightly alkaline.

2. In alkaline solution, the zwitterion form of glycine would tend to lose
    a proton and become the basic form:

    $$CH_3NH_3{}^+CO_2{}^-(aq) + OH^-(aq) \rightleftharpoons CH_2NH_2CO_2{}^-(aq) + H_2O(l)$$

3. The basic form of glycine can make two bonds, using the lone pair of
    electrons on the nitrogen and the negative charge on the carboxyl oxygen.

$$CH_2NH_2C{\overset{\displaystyle O}{\underset{\displaystyle O^-}{}}}$$

Glycine is therefore a bidentate ligand.

Experiment 97. Questions

1. The intensity of the purple (or pink) colour is related to the concen-
    tration of the protein. A colorimeter or spectrophotometer could be
    calibrated to enable concentration to be determined from colour intensity.

2.
$$
\begin{array}{ccc}
O=C & & C=O \\
| & & | \\
NH & & NH \\
R-CH & & CH-R \\
| & Cu & | \\
O=C & & C=O \\
NH & & NH \\
| & & | \\
R-CH & & CH-R \\
\end{array}
$$

3. No, because the biuret complex is formed from nitrogen in peptide bonds.
    However, care is needed because some amino acids form bluish complexes
    with copper(II) ions (see Experiment 96) and the colours can be confused.

Experiment 98. Specimen results

Results Table 98

| | Distances travelled/cm | | |
Amino acid	by solvent	by amino acid	R_f value
Aspartic acid - alone	7.1	1.1	0.15
- in mixture	6.8	0.9	0.13
Leucine - alone	7.0	5.5	0.79
- in mixture	6.8	5.1	0.75
Lysine - alone	7.1	2.6	0.37
- in mixture	6.8	2.4	0.35

Experiment 98. Questions

1. R_f values vary slightly with local conditions such as temperature, the
    purity of the solvent, the nature of the chromatography paper, air-
    pressure and humidity.

2. The greater the solubility of a substance in a particular solvent, the
    less readily it is retained by the stationary phase and the faster it
    appears to move. This gives a higher R_f value.

3. Amino acids are readily transferred from moist and/or dirty fingers on
    to the paper and show up as confusing brown marks when sprayed with
    ninhydrin solution.

Results Table 99a

	Glucose	Fructose	Sucrose	Maltose	Starch
A. Dehydration					
(a) Action of heat	Melted easily to a clear liquid which became dark & viscous. Steamy gas.		Melted easily to a clear liquid which became dark & viscous. Steamy gas		
(b) Sulphuric acid	Mixture turned black & swelled giving off heat & much steam. Crumbly black solid left. CO_2 & SO_2 detected.		Mixture turned black & swelled giving off heat & much steam. Crumbly black solid left. CO_2 & SO_2 detected.		
B. Oxidation					
(a) Fehling's soln.	Reddish ppt.	Reddish ppt.	No change.	Reddish ppt.	
(b) Tollens' reagent	Silver mirror.	Silver mirror.	No change.	Silver mirror.	
C. Carbonyl derivatives					
(a) 2,4-dinitro-phenylhydrazine	Yellow ppt. (slow)	Yellow ppt. (slow)			
(b) phenylhydrazine	Yellow ppt. 5 mins M.Pt. 205 °C	Yellow ppt. 3 mins M.Pt. 205 °C			
D. Hydrolysis				Red ppt. with Fehling's solution.	A & C. Complete hydrolysis to reducing sugar. No starch left. B. Partial hydrolysis.

Experiment 99. Specimen results

Results Table 99b

	Movement of analyzer from zero			
Glucose	Initial $+27°$	$\frac{1}{2}$ volume $+13°$	Diluted $+13°$	+ indicates clockwise − indicates anticlockwise

Sucrose	Time, t/min	0	3	6	10	15	20	30	∞
	Analyzer reading, $\alpha_t/°$	+34	+18	+ 7	− 1	− 8	− 9	−11	−12
	$\alpha_t - \alpha_\infty/°$	46	30	19	11	4	3	1	0

Experiment 99. Questions

1. The main products are carbon and water.

$$C_6H_{12}O_6 \; \rightarrow \; 6C + 6H_2O$$

2. The reaction is called dehydration because water molecules are removed.

3. Some of the sulphuric acid is reduced by the hot carbon to sulphur
    dioxide, while some of the carbon is oxidized to carbon dioxide.

$$C(s) + 2H_2SO_4(l) \; \rightarrow \; CO_2(g) + 2SO_2(g) + 2H_2O(l)$$

    Some is merely diluted by the water released.

4. In each case, glucose is oxidized to gluconic acid (the $-$CHO group
    becomes $-CO_2H$). Tollens' reagent is reduced to metallic silver, and
    Fehling's solution is reduced to copper(I) oxide.

5. Glucose reduces Tollens' reagent and Fehling's solution because it has
    an aldehyde group $-$CHO. Fructose is also a reducing sugar, even though
    it has no $-$CHO group; the combination $-C-C-$ is an effective
    reductant.

    In sucrose, however, the link between the glucose and fructose units is
    made via the reducing groups which then become ineffective.

6. The product is called glucosazone.

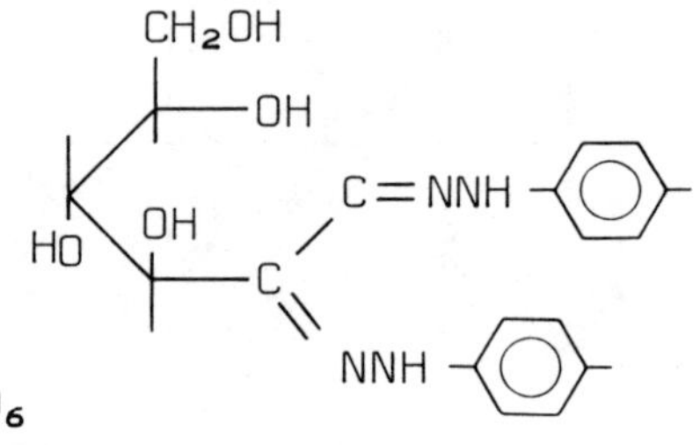

7. $C_{12}H_{22}O_{11} + H_2O \; \rightarrow \; C_6H_{12}O_6 + C_6H_{12}O_6$
    sucrose   glucose   fructose

8. Both sucrose and glucose are dextrorotatory; fructose is laevorotatory
    to a greater extent than glucose is dextrorotatory. The final mixture
    is therefore laevorotatory. 'Inversion' refers to the change from
    dextrorotation to laevorotation, (+) to (−).

9. The inversion of sucrose is catalyzed by invertase.

10. $[\alpha] = \dfrac{\alpha}{lc} \qquad \therefore \; c = \dfrac{\alpha}{[\alpha]l} = \dfrac{27°}{52.5° \; g^{-1} \; cm^3 \; dm^{-1} \times 1 \; dm} = 0.51 \; g \; cm^{-3}$

11. $\alpha_t - \alpha_\infty$ is proportional to the amount of sucrose present in the solution.
    The slope of the curve is therefore the rate of the reaction and this
    decreases to zero as the reaction proceeds to completion. (This is a
    good example of a first order reaction.)

ENLARGED RESULTS TABLES TO ACCOMPANY THE OBSERVATION AND DEDUCTION EXERCISES

Results Table 60a

	Test	Observations	Inferences
(a)	Heat approximately 0.1 g of F in a pyrex tube, at first gently and then more strongly, until the change is complete. Cool and keep the residue. Test any gases evolved.		
(b)	Make an aqueous solution of the residue from (a) and carry out the following tests on portions: (i) Add aqueous silver nitrate followed by dilute nitric acid. (ii) Add aqueous chlorine. (iii) Add aqueous lead(II) ethanoate (lead acetate).		

Results Table 60b

Substance provided	Name of non-metal	Name and formula of substance/ion	Oxidation number
F			

Results Table 61a   Tests with unknown substance D

Method	Observations	Inferences

Results Table 61c   Experiments to test inferences

Inference tested	Test and observations	Conclusion

Results Table 61b Tests with unknown substance E

Method	Observations	Inferences
Method	Observations	Inferences

Method	Observation	Inference
(1) A spatula-ful of the powder was heated in an ignition tube, gently at first and then more strongly. A glowing splint was lowered into the tube.		
(2) About 2 cm^3 of dilute nitric acid was added to half a spatula-ful of the powder in a test-tube. The mixture was gently heated.		
(3) ·The colourless liquid from test 2 was decanted off carefully into a test-tube. (a) About 1 cm^3 was poured into another test-tube and a few drops of KI solution were added. (b) NaOH(aq) was added to half the remaining solution from (2), drop by drop initially and then to excess.		

Results Table 67b   Experiment to test inferences

Inference	Test and observation	Conclusion

Results Table 67c

Test	Observations	Inferences
1. To 1 cm³ of the solution of H add aqueous silver nitrate followed by dilute nitric acid.		
2. To 1 cm³ of the solution of H add aqueous sodium hydroxide until in excess.		
3. To 1 cm³ of aqueous iron(III) chloride add a few drops of aqueous potassium thiocyanate. To this solution add some of the solution of H.		
4. To 1 cm³ of aqueous mercury(II) chloride add a little of the solution of H, then excess.		
5. Heat some of I in a pyrex boiling tube. Allow to cool. Add 8-10 cm³ of dilute nitric acid to the residue and boil the mixture for 1 or 2 minutes. Filter if necessary and use portions of the cool solution for the following tests: (a) To 1 cm³ of the solution add aqueous sodium hydroxide. (b) To 1 cm³ of the solution add dilute sulphuric acid. (c) To 1 cm³ of the solution add aqueous potassium chromate(VI).		

Results Table 75

Test	Observations	Inferences
1. Warm three-quarters of your sample of C with 4-5 cm³ of aqueous sodium hydroxide.		
2. Add approximately 10 cm³ of dilute sulphuric acid to the remainder of your sample of C in a boiling-tube and warm the mixture in order to dissolve the solid. Use portions of the solution for the following tests:		
(a) To 1-2 cm³ of the solution of C add an equal volume of aqueous potassium iodide. Then add, dropwise, aqueous sodium thiosulphate until there is no further change.		
(b) To 2 cm³ of the solution add a little copper powder and warm.		
(c) To 3-4 cm³ of the solution add a little zinc powder and allow the mixture to stand. It is suggested that you make observations for about 5 minutes and then allow the mixture to stand for a further 30 minutes, making observation from time to time. During this time you should proceed with other tests.		
3. Carry out a flame test on substance D.		
4. Dissolve some D in the minimum quantity of distilled water and use portions of the solution for the following tests:		
(a) Add a few drops of this solution to a mixture of 1-2 cm³ of aqueous potassium iodide with an equal volume of dilute sulphuric acid. Then add aqueous sodium thiosulphate dropwise.		
(b) To 1 cm³ of the solution of D add aqueous sodium hydroxide. Then add dilute sulphuric acid until in excess.		
(c) To about 2 cm³ of the solution of D add an equal volume of dilute sulphuric acid. Then add hydrogen peroxide solution dropwise until there is no further change.		(No inference required)
(d) Transfer the solution from (c) to a boiling-tube and add about 10 cm³ of aqueous sodium hydroxide and 1 cm³ of hydrogen peroxide solution. Heat the resulting solution.		

Results Table 80

Test	Observations	Inferences
1. To 2 or 3 cm³ of solution B add a few drops of dilute hydrochloric acid. Now add a few drops of aqueous barium chloride.		
2. To 2 or 3 cm³ of solution B add a few drops of aqueous lead ethanoate (lead acetate). Now add an excess of aqueous ammonium ethanoate (ammonium acetate) and shake the mixture.		
3. To about 5 cm³ of solution C in a boiling-tube add about 10 cm³ of dilute hydrochloric acid. Allow the mixture to stand for a minute or two. Cautiously smell the mixture. Warm, if necessary, and test the gas evolved. Describe below how you performed this test.		
4. (a) In a small beaker place 2 or 3 cm³ of aqueous copper(II) sulphate. Acidify with two drops of dilute sulphuric acid. Now add about 5 cm³ of aqueous potassium iodide. (b) To the mixture obtained in 4(a) add solution C until in excess, swirling well.		
5. (a) In a small beaker mix about 2 cm³ of aqueous sodium chloride with an equal volume of aqueous silver nitrate. (b) To the mixture obtained in 5(a) add solution C until in excess, swirling well.		
6. Dissolve about 1 g of iron(II) sulphate crystals in dilute sulphuric acid and add a few drops of aqueous potassium thiocyanate. Now add a few drops of solution D.		
7. (a) In a small beaker place about 5 cm³ of aqueous potassium iodide and a few drops of dilute sulphuric acid. Now add about 5 cm³ of solution D. Warm this mixture. (b) To the mixture obtained in 7(a) add solution C until in excess, swirling well.		
8. Evaporate a small quantity of solution B to dryness and perform a flame test on the residue. Describe below how you do this.		

Test	Observations	Inferences
1.(a) Place 1 cm³ of C in a test-tube and add an equal volume of water. Now add a little anhydrous sodium carbonate.		
1.(b) Repeat test 1.(a) using D.		
2.(a) (The reactions 2.(a) and 2.(b) should be performed in a fume cupboard.) Place 2 or 3 cm³ of C in a dry beaker. Add a little phosphorus pentachloride. (CARE!)		
2.(b) Repeat test 2.(a) using D. (CARE!)		
3. Mix, in a small beaker, about 2 cm³ of each of C, D and concentrated sulphuric acid. (CARE!) Warm gently but do not boil. Pour the mixture into an excess of aqueous sodium carbonate in a large beaker. Smell the product.		
4. Mix about 5 cm³ of D with an equal volume of aqueous potassium dichromate in a test-tube. Pour the mixture into about 10 cm³ of nearly boiling dilute sulphuric acid in a small beaker. Smell the mixture.		
5. In a small beaker dissolve a few crystals of potassium iodide in about 10 cm³ of D. Add a few drops of aqueous sodium hypochlorite. Warm the mixture but do not boil. Cautiously smell the products.		
6. Make a solution of E in distilled water and use portions for the following tests: To 2-3 cm³ of the solution add aqueous silver nitrate. Then add dilute nitric acid. Finally add dilute aqueous ammonia.		
7. Dissolve a little E in about 1 cm³ of concentrated hydrochloric acid and dilute to about 4 cm³ with distilled water. Cool the tube in an ice-water mixture and add a few drops of aqueous sodium nitrite (to be prepared by dissolving sodium nitrite in distilled water). Leave the tube in the ice-water mixture. Dissolve a few crystals of phenol (CARE!) in 7-8 cm³ of aqueous sodium hydroxide, cool this solution, and then add it to the cold solution prepared as above.		

Results Table 92

Test	Observations	Inferences

(a) Dissolve some of I in the minimum quantity of distilled water and use portions of the solution for the following tests.
 	
   (i) To 1-2 cm^3 of the solution of I add aqueous sodium hydroxide until in excess.
 	
   (ii) To 1-2 cm^3 of the solution of I add an equal volume of dilute sulphuric acid, shake the tube, and allow the solution to stand.		

(b) Add a little dilute sulphuric acid to some solid I and heat the mixture.		
(c) Heat a little solid I in a pyrex test-tube.		
(d) Using the apparatus shown, heat a little of I in a test-tube and pass the vapour evolved into 2,4-dinitrophenylhydrazine reagent.		
(e) Heat a little solid J in a pyrex test-tube.		
(f) To a little solid J add a little concentrated sulphuric acid (CARE!) Warm gently.		

Results Table 95

EXPERIMENT	OBSERVATIONS	INFERENCES
(a) To 2 cm^3 of the solution of F add 2, 4-dinitrophenylhydrazine reagent.		
(b) Prepare a sample of Tollens' reagent as follows: to 5 cm^3 of aqueous silver nitrate in a test-tube add 1-2 drops of aqueous sodium hydroxide. Then add dilute aqueous ammonia until only a trace of precipitate remains. Now add 5 drops of the solution of F and place the tube in hot water. (Pour the contents of the tube down the sink on completion of the test.)		
(c) Add a little of the solution of F to some sodium hydrogencarbonate.		
(d) To 1 cm^3 of the solution of F add 3 cm^3 of aqueous potassium iodide and then 10 cm^3 of sodium chlorate(I) (sodium hypochlorite) solution.		
(e) Place ONE drop of G on an inverted crucible lid and ignite G from above.		
(f) To 1-2 cm^3 of G add 2,4-dinitro-phenylhydrazine reagent.		
(g) Prepare a sample of Tollens' reagent as in (b). Add 5 drops of G, shake the mixture and place the tube in hot water. (N.B. Pour the contents of the tube down the sink on completion of the test.)		
(h) To 0.5 cm^3 of G add 2 cm^3 of 6 M sodium hydroxide. Warm the mixture and stir well for 5 minutes. Then add sufficient water to dissolve any residue which has formed. Separate a portion of the aqueous layer and add a little concentrated hydrochloric acid to it. Cool.		
(i) Test to see if H is miscible with (i) cold water, (ii) hot water.		
(j) (i) To about 2 cm^3 of H in a dry beaker add a little anhydrous sodium carbonate. (ii) Now add about 5 cm^3 of hot water to the mixture from test (j),(i).		
(k) (This reaction should be performed in a fume cupboard.) Place 2 or 3 cm^3 of H in a dry beaker. Add a little phosphorus pentachloride (CARE!).		
(l) Place 2 cm^3 of glacial ethanoic acid and 2 cm^3 of H in a boiling tube and add 2 cm^3 of phenylamine (aniline). Heat cautiously until the mixture begins to boil. Remove from the flame and allow to stand for 2 minutes. Now pour the mixture into about 125 cm^3 of cold water with stirring and allow to stand.		